CliffsNotes®
6th Grade Common Core Math

2nd Edition

Course Hero

CliffsNotes® 6th Grade Common Core Math

2nd Edition

By Sandra Luna McCune, Ph.D.

Course Hero

About the Author

Sandra Luna McCune, Ph.D., is professor emeritus and a former Regents professor in the Department of Elementary Education at Stephen F. Austin State University, where she received the Distinguished Professor Award. She now is a full-time author and consultant and resides near Austin, Texas.

Acknowledgments

I would like to thank Grace Freedson, Greg Tubach, and Christina Stambaugh for their support and encouragement during completion of this book. I also owe a debt of gratitude to Lynn Northrup, Tom Page, and Mary Jane Sterling for their meticulous editing and invaluable suggestions and to Donna Wright for her careful proofing of the manuscript.

Dedication

This book is dedicated to my grandchildren—Richard, Rose, Jude, Sophia, Josephine, and Myla Mae. They fill my life with joy!

Editorial

Executive Editor: Greg Tubach
Senior Editor: Christina Stambaugh
Copy Editor: Lynn Northrup
Technical Editors: Tom Page and Mary Jane Sterling
Proofreader: Donna Wright

CliffsNotes® 6th Grade Common Core Math: Quick Review (2nd Edition)

Library of Congress Control Number: 2014960089
ISBN: 978-1-957671-06-2

Printed in the United States of America
DOC 10 9 8 7 6 5 4 3 2 1

Table of Contents

Introduction

This book is organized around the Grade 6 Common Core State Standards for Mathematics. These standards define what sixth-grade students are expected to understand and be able to do in their study of mathematics. They include content standards and mathematical practice standards.

In Grade 6, the content standards are grouped under five domains:

- Ratios and Proportional Relationships
- The Number System
- Expressions and Equations
- Geometry
- Statistics and Probability

Ratios and Proportional Relationships

Understand ratio concepts and use ratio reasoning to solve problems.

CCSS.Math.Content.6.RP.A.1 Understand the concept of a ratio and use ratio language to describe a ratio relationship between two quantities. *For example, "The ratio of wings to beaks in the bird house at the zoo was 2:1, because for every 2 wings there was 1 beak." "For every vote candidate A received, candidate C received nearly three votes."*

CCSS.Math.Content.6.RP.A.2 Understand the concept of a unit rate $\frac{a}{b}$ associated with a ratio $a:b$ with $b \neq 0$, and use rate language in the context of a ratio relationship. *For example, "This recipe has a ratio of 3 cups of flour to 4 cups of sugar, so there is* $\frac{3}{4}$ *cup of flour for each cup of sugar." "We paid \$75 for 15 hamburgers, which is a rate of \$5 per hamburger."*

CCSS.Math.Content.6.RP.A.3 Use ratio and rate reasoning to solve real-world and mathematical problems, e.g., by reasoning about tables of equivalent ratios, tape diagrams, double number line diagrams, or equations.

- **CCSS.Math.Content.6.RP.A.3.A** Make tables of equivalent ratios relating quantities with whole-number measurements, find missing values in the tables, and plot the pairs of values on the coordinate plane. Use tables to compare ratios.
- **CCSS.Math.Content.6.RP.A.3.B** Solve unit rate problems including those involving unit pricing and constant speed. *For example, if it took 7 hours to mow 4 lawns, then at that rate, how many lawns could be mowed in 35 hours? At what rate were lawns being mowed?*
- **CCSS.Math.Content.6.RP.A.3.C** Find a percent of a quantity as a rate per 100 (e.g., 30% of a quantity means $\frac{30}{100}$ times the quantity); solve problems involving finding the whole, given a part and the percent.
- **CCSS.Math.Content.6.RP.A.3.D** Use ratio reasoning to convert measurement units; manipulate and transform units appropriately when multiplying or dividing quantities.

The Number System

Apply and extend previous understandings of multiplication and division to divide fractions by fractions.

CCSS.Math.Content.6.NS.A.1 Interpret and compute quotients of fractions, and solve word problems involving division of fractions by fractions, e.g., by using visual fraction models and equations to represent the problem. *For example, create a story context for* $\frac{2}{3} \div \frac{3}{4}$ *and use a visual fraction model to show the quotient; use the relationship between multiplication and division to explain that* $\frac{2}{3} \div \frac{3}{4} = \frac{8}{9}$ *because* $\frac{3}{4}$ *of* $\frac{8}{9}$ *is* $\frac{2}{3}$. *(In general,* $\frac{a}{b} \div \frac{c}{d} = \frac{ad}{bc}$.*) How much chocolate will each person get if 3 people share* $\frac{1}{2}$ *lb of chocolate equally? How many* $\frac{3}{4}$*-cup servings are in* $\frac{2}{3}$ *of a cup of yogurt? How wide is a rectangular strip of land with length* $\frac{3}{4}$ *mi and area* $\frac{1}{2}$ *square mi?*

Compute fluently with multi-digit numbers and find common factors and multiples.

CCSS.Math.Content.6.NS.B.2 Fluently divide multi-digit numbers using the standard algorithm.

CCSS.Math.Content.6.NS.B.3 Fluently add, subtract, multiply, and divide multi-digit decimals using the standard algorithm for each operation.

CCSS.Math.Content.6.NS.B.4 Find the greatest common factor of two whole numbers less than or equal to 100 and the least common multiple of two whole numbers less than or equal to 12. Use the distributive property to express a sum of two whole numbers 1–100 with a common factor as a multiple of a sum of two whole numbers with no common factor. *For example, express 36 + 8 as 4(9 + 2).*

Apply and extend previous understandings of numbers to the system of rational numbers.

CCSS.Math.Content.6.NS.C.5 Understand that positive and negative numbers are used together to describe quantities having opposite directions or values (e.g., temperature above/below zero, elevation above/below sea level, credits/debits, positive/negative electric charge); use positive and negative numbers to represent quantities in real-world contexts, explaining the meaning of 0 in each situation.

CCSS.Math.Content.6.NS.C.6 Understand a rational number as a point on the number line. Extend number line diagrams and coordinate axes familiar from previous grades to represent points on the line and in the plane with negative number coordinates.

- **CCSS.Math.Content.6.NS.C.6.A** Recognize opposite signs of numbers as indicating locations on opposite sides of 0 on the number line; recognize that the opposite of the opposite of a number is the number itself, e.g., – (–3) = 3, and that 0 is its own opposite.
- **CCSS.Math.Content.6.NS.C.6.B** Understand signs of numbers in ordered pairs as indicating locations in quadrants of the coordinate plane; recognize that when two ordered pairs differ only by signs, the locations of the points are related by reflections across one or both axes.
- **CCSS.Math.Content.6.NS.C.6.C** Find and position integers and other rational numbers on a horizontal or vertical number line diagram; find and position pairs of integers and other rational numbers on a coordinate plane.

CCSS.Math.Content.6.NS.C.7 Understand ordering and absolute value of rational numbers.

- **CCSS.Math.Content.6.NS.C.7.A** Interpret statements of inequality as statements about the relative position of two numbers on a number line diagram. *For example, interpret –3 > –7 as a statement that –3 is located to the right of –7 on a number line oriented from left to right.*
- **CCSS.Math.Content.6.NS.C.7.B** Write, interpret, and explain statements of order for rational numbers in real-world contexts. *For example, write –3°C > –7°C to express the fact that –3°C is warmer than –7°C.*
- **CCSS.Math.Content.6.NS.C.7.C** Understand the absolute value of a rational number as its distance from 0 on the number line; interpret absolute value as magnitude for a positive or negative quantity in a real-world situation. *For example, for an account balance of –30 dollars, write |–30| = 30 to describe the size of the debt in dollars.*
- **CCSS.Math.Content.6.NS.C.7.D** Distinguish comparisons of absolute value from statements about order. *For example, recognize that an account balance less than –30 dollars represents a debt greater than 30 dollars.*

CCSS.Math.Content.6.NS.C.8 Solve real-world and mathematical problems by graphing points in all four quadrants of the coordinate plane. Include use of coordinates and absolute value to find distances between points with the same first coordinate or the same second coordinate.

Expressions and Equations

Apply and extend previous understandings of arithmetic to algebraic expressions.

CCSS.Math.Content.6.EE.A.1 Write and evaluate numerical expressions involving whole-number exponents.

CCSS.Math.Content.6.EE.A.2 Write, read, and evaluate expressions in which letters stand for numbers.

- **CCSS.Math.Content.6.EE.A.2.A** Write expressions that record operations with numbers and with letters standing for numbers. *For example, express the calculation "Subtract y from 5" as* $5 - y$.
- **CCSS.Math.Content.6.EE.A.2.B** Identify parts of an expression using mathematical terms (sum, term, product, factor, quotient, coefficient); view one or more parts of an expression as a single entity. *For example, describe the expression* $2(8 + 7)$ *as a product of two factors; view* $(8 + 7)$ *as both a single entity and a sum of two terms.*
- **CCSS.Math.Content.6.EE.A.2.C** Evaluate expressions at specific values of their variables. Include expressions that arise from formulas used in real-world problems. Perform arithmetic operations, including those involving whole-number exponents, in the conventional order when there are no parentheses to specify a particular order (Order of Operations). *For example, use the formulas* $V = s^3$ *and* $A = 6s^2$ *to find the volume and surface area of a cube with sides of length* $s = \frac{1}{2}$.

CCSS.Math.Content.6.EE.A.3 Apply the properties of operations to generate equivalent expressions. *For example, apply the distributive property to the expression* $3(2 + x)$ *to produce the equivalent expression* $6 + 3x$*; apply the distributive property to the expression* $24x + 18y$ *to produce the equivalent expression* $6(4x + 3y)$*; apply properties of operations to* $y + y + y$ *to produce the equivalent expression* $3y$.

CCSS.Math.Content.6.EE.A.4 Identify when two expressions are equivalent (i.e., when the two expressions name the same number regardless of which value is substituted into them). *For example, the expressions* $y + y + y$ *and* $3y$ *are equivalent because they name the same number regardless of which number* y *stands for.*

Reason about and solve one-variable equations and inequalities.

CCSS.Math.Content.6.EE.B.5 Understand solving an equation or inequality as a process of answering a question: Which values from a specified set, if any, make the equation or inequality true? Use substitution to determine whether a given number in a specified set makes an equation or inequality true.

CCSS.Math.Content.6.EE.B.6 Use variables to represent numbers and write expressions when solving a real-world or mathematical problem; understand that a variable can represent an unknown number, or, depending on the purpose at hand, any number in a specified set.

CCSS.Math.Content.6.EE.B.7 Solve real-world and mathematical problems by writing and solving equations of the form $x + p = q$ and $px = q$ for cases in which p, q, and x are all nonnegative rational numbers.

CCSS.Math.Content.6.EE.B.8 Write an inequality of the form $x > c$ or $x < c$ to represent a constraint or condition in a real-world or mathematical problem. Recognize that inequalities of the form $x > c$ or $x < c$ have infinitely many solutions; represent solutions of such inequalities on number line diagrams.

Represent and analyze quantitative relationships between dependent and independent variables.

CCSS.Math.Content.6.EE.C.9 Use variables to represent two quantities in a real-world problem that change in relationship to one another; write an equation to express one quantity, thought of as the dependent variable, in terms of the other quantity, thought of as the independent variable. Analyze the relationship between the dependent and independent variables using graphs and tables, and relate these to the equation. *For example, in a problem involving motion at constant speed, list and graph ordered pairs of distances and times, and write the equation $d = 65t$ to represent the relationship between distance and time.*

Geometry

Solve real-world and mathematical problems involving area, surface area, and volume.

CCSS.Math.Content.6.G.A.1 Find the area of right triangles, other triangles, special quadrilaterals, and polygons by composing into rectangles or decomposing into triangles and other shapes; apply these techniques in the context of solving real-world and mathematical problems.

CCSS.Math.Content.6.G.A.2 Find the volume of a right rectangular prism with fractional edge lengths by packing it with unit cubes of the appropriate unit fraction edge lengths, and show that the volume is the same as would be found by multiplying the edge lengths of the prism. Apply the formulas $V = lwh$ and $V = Bh$ to find volumes of right rectangular prisms with fractional edge lengths in the context of solving real-world and mathematical problems.

CCSS.Math.Content.6.G.A.3 Draw polygons in the coordinate plane given coordinates for the vertices; use coordinates to find the length of a side joining points with the same first coordinate or the same second coordinate. Apply these techniques in the context of solving real-world and mathematical problems.

CCSS.Math.Content.6.G.A.4 Represent three-dimensional figures using nets made up of rectangles and triangles, and use the nets to find the surface area of these figures. Apply these techniques in the context of solving real-world and mathematical problems.

Statistics and Probability

Develop understanding of statistical variability.

CCSS.Math.Content.6.SP.A.1 Recognize a statistical question as one that anticipates variability in the data related to the question and accounts for it in the answers. *For example, "How old am I?" is not a statistical question, but "How old are the students in my school?" is a statistical question because one anticipates variability in students' ages.*

CCSS.Math.Content.6.SP.A.2 Understand that a set of data collected to answer a statistical question has a distribution which can be described by its center, spread, and overall shape.

CCSS.Math.Content.6.SP.A.3 Recognize that a measure of center for a numerical data set summarizes all of its values with a single number, while a measure of variation describes how its values vary with a single number.

Summarize and describe distributions.

CCSS.Math.Content.6.SP.B.4 Display numerical data in plots on a number line, including dot plots, histograms, and box plots.

CCSS.Math.Content.6.SP.B.5 Summarize numerical data sets in relation to their context, such as by:

- **CCSS.Math.Content.6.SP.B.5.A** Reporting the number of observations.
- **CCSS.Math.Content.6.SP.B.5.B** Describing the nature of the attribute under investigation, including how it was measured and its units of measurement.
- **CCSS.Math.Content.6.SP.B.5.C** Giving quantitative measures of center (median and/or mean) and variability (interquartile range and/or mean absolute deviation), as well as describing any overall pattern and any striking deviations from the overall pattern with reference to the context in which the data were gathered.
- **CCSS.Math.Content.6.SP.B.5.D** Relating the choice of measures of center and variability to the shape of the data distribution and the context in which the data were gathered.

Mathematical Practice

The Standards for Mathematical Practice describe the ways in which the students ought to engage with the mathematics content standards as they develop in mathematical proficiency and understanding.

CCSS.Math.Practice.MP.1 Make sense of problems and persevere in solving them.

CCSS.Math.Practice.MP.2 Reason abstractly and quantitatively.

CCSS.Math.Practice.MP.3 Construct viable arguments and critique the reasoning of others.

CCSS.Math.Practice.MP.4 Model with mathematics.

CCSS.Math.Practice.MP.5 Use appropriate tools strategically.

CCSS.Math.Practice.MP.6 Attend to precision.

CCSS.Math.Practice.MP.7 Look for and make use of structure.

CCSS.Math.Practice.MP.8 Look for and express regularity in repeated reasoning.

Mathematical Practice

1. Ratios and Proportional Relationships

In this chapter, you will learn about ratios, rates, unit rates, equivalent ratios and their representations, and percents. You will use ratio and rate reasoning and proportional relationships to solve real-world and mathematical problems.

Understanding the Concept of a Ratio

(CCSS.Math.Content.6.RP.A.1)

A **ratio** is the result of a multiplicative comparison of two quantities or measures. For example, in a paint mixture that uses three parts white paint to four parts blue paint, the ratio of "white paint to blue paint" is three to four. For every three parts of white paint, you need four parts of blue paint. Thus, for every 3 pints of white paint, you need 4 pints of blue paint; for every 3 quarts of white paint, you need 4 quarts of blue paint, and so on. You can express the ratio "three to four" in three different forms. You can write the ratio as 3 to 4, 3:4, or $\frac{3}{4}$. (***Tip:*** The colon in "3:4" is read "to.") The numbers 3 and 4 are the **terms** of the ratio. The terms are nonnegative numbers, and the ratio does not have any units. You can represent the ratio with a **tape diagram** like this.

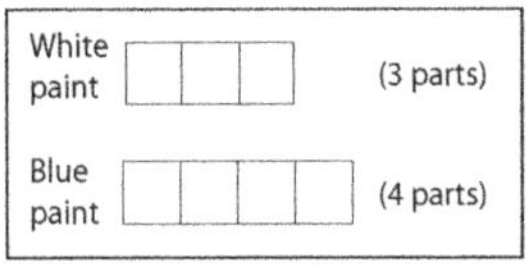

Notice the order of the numbers in a ratio is important. In the paint mixture, the ratio of "blue paint to white paint" is four to three; that is, 4 to 3, 4:3, or $\frac{4}{3}$. The description of the ratio relationship gives you the order of the numbers in the ratio. The ratio of "a items to b items" is a to b, a:b, or $\frac{a}{b}$.

Ratios can be part-to-part comparisons or part-to-whole comparisons. The paint mixture example is a part-to-part comparison. The part of the mixture that is white paint is compared to the part of the mixture that is blue paint to obtain the ratio 3 to 4. The ratio of the number of girls in a club to the number of members in the club is a part-to-whole ratio. If there are 7 girls in a club of 10 members, the ratio of girls (the part) to members (the whole) is 7 to 10.

☞Try These

1. The ratio of the number of girls to the number of boys in a club is 7 to 3. Create a tape diagram to represent the ratio.

2. A farmer has 10 sheep and 25 chickens. Write ratios for the following descriptions. Write your answers in three forms.
   (a) the ratio of the number of sheep to the number of chickens
   (b) the ratio of the number of chickens to the number of sheep
   (c) the ratio of the number of sheep to the total number of sheep and chickens
   (d) the ratio of the number of chickens to the total number of sheep and chickens

3.

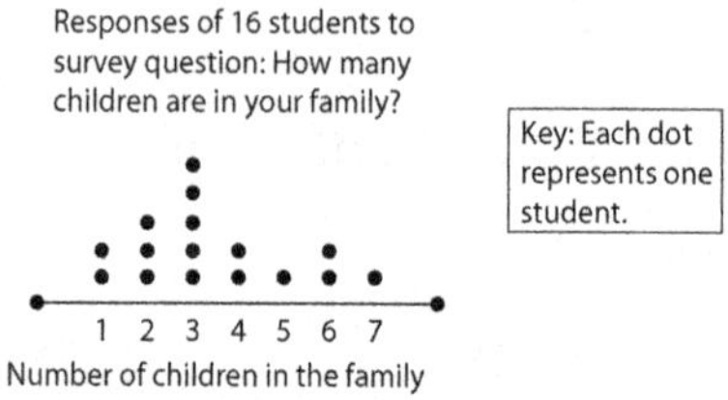

Based on the data shown in the preceding dot plot, write ratios for the following descriptions. Write your answers in three forms.
   (a) the ratio of the number of students with one child in the family to the number of students with seven children in the family
   (b) the ratio of the number of students with fewer than four children in the family to the number of students with four or more children in the family
   (c) the ratio of the number of students with fewer than four children in the family to the total number of students surveyed
   (d) the ratio of the number of students with four or more children in the family to the total number of students surveyed

Solutions

1.

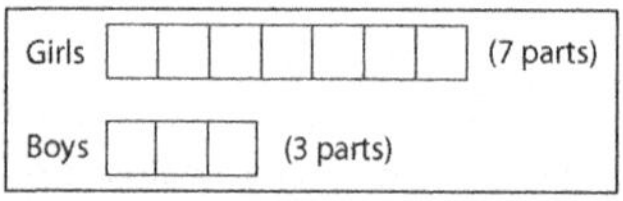

2. **(a)** The ratio of the number of sheep to the number of chickens is 10 to 25, 10:25, or $\frac{10}{25}$.

   (b) The ratio of the number of chickens to the number of sheep is 25 to 10, 25:10, or $\frac{25}{10}$.
   (c) Make a table to show the information in the question.

Number of Sheep	Number of Chickens	Total
10	25	35

   The total number of sheep and chickens is $10 + 25 = 35$. The ratio of the number of sheep to the total is 10 to 35, 10:35, or $\frac{10}{35}$.
   (d) Refer to the table in part (c). The total number of sheep and chickens is $10 + 25 = 35$. The ratio of the number of chickens to the total is 25 to 35, 25:35, or $\frac{25}{35}$.

3. **(a)** The ratio of the number of students with one child in the family to the number of students with seven children in the family is 2 to 1, 2:1, or $\frac{2}{1}$.

   (b) The ratio of the number of students with fewer than four children in the family to the number of students with four or more children in the family is 10 to 6, 10:6, or $\frac{10}{6}$.

   (c) The ratio of the number of students with fewer than four children in the family to the total number of students surveyed is 10 to 16, 10:16, or $\frac{10}{16}$.

   (d) The ratio of the number of students with four or more children in the family to the total number of students surveyed is 6 to 16, 6:16, or $\frac{6}{16}$.

Generating and Using Equivalent Ratios

(CCSS.Math.Content.6.RP.A.1, CCSS.Math.Content.6.RP.A.3.A)

Two ratios are **equivalent** if there is a positive number that can be multiplied by both terms in one ratio to equal the corresponding terms in the second ratio. That is, ratio $\frac{a}{b}$ equals ratio $\frac{c}{d}$ if there is a positive number k such that $c = a \times k$ and $d = b \times k$. Here are examples.

Ratio $\frac{7}{3}$ equals ratio $\frac{14}{6}$ because there is a positive number 2 such that $\frac{7\times 2}{3\times 2} = \frac{14}{6}$.

Ratio $\frac{2}{7}$ equals ratio $\frac{10}{35}$ because there is a positive number 5 such that $\frac{2\times 5}{7\times 5} = \frac{10}{35}$.

You can use tape diagrams to generate equivalent ratios. Suppose the ratio of the number of girls to the number of boys in a student club is 7 to 3. Three equivalent ratios for the number of girls to the number of boys in the club are shown in the following tape diagrams.

Each equal-sized part of the tape represents one student. The ratio is 7 to 3:

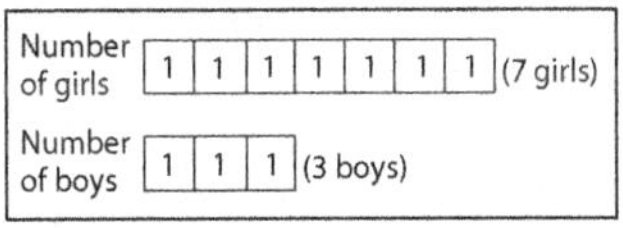

Each equal-sized part of the tape represents two students. The ratio is 14 to 6. This ratio is equivalent to 7 to 3 because $14 = 7 \times 2$ and $6 = 3 \times 2$:

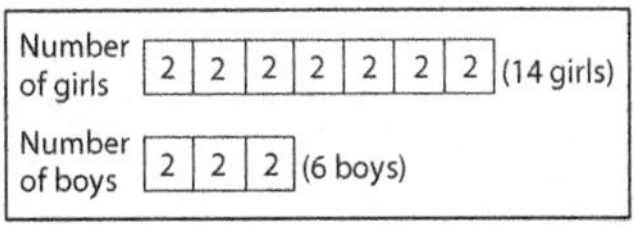

Each equal-sized part of the tape represents three students. The ratio is 21 to 9. This ratio is equivalent to 7 to 3 because $21 = 7 \times 3$ and $9 = 3 \times 3$:

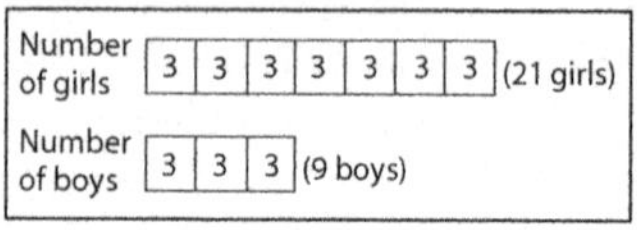

A **ratio table** lists rows of equivalent ratios. Suppose a farmer has 2 sheep for every 7 chickens. Here is a ratio table with five rows of equivalent ratios for this relationship.

Number of Sheep	Number of Chickens
2	7
4	14
6	21
8	28
10	35

You generate the rows of equivalent ratios by multiplying the terms 2 and 7 by the same number. To obtain the second row, multiply both terms by 2. To obtain the third row, multiply both terms by 3. To obtain the fourth row, multiply both terms by 4. And to obtain the fifth row, multiply both terms by 5. It's okay if you want to skip rows. To obtain the tenth row, multiply both terms by 10. The result is 20 sheep to 70 chickens.

☞ Try These

1. The ratio of the number of apples Arianna has to the number of apples Izel has is 2 to 3. Arianna has 8 apples. Use a tape diagram to determine how many apples Izel has.
   - **(a)** Make a tape diagram that represents the ratio 2 to 3.
   - **(b)** Arianna has 8 apples. How many apples does each equal-sized part of the tape diagram in part (a) represent?
   - **(c)** How many apples does Izel have?
2. The ratio of the number of adults to children at the park is 5 to 7. The total number of people at the park is 108. Use a tape diagram to determine the number of children at the park.
   - **(a)** Make a tape diagram that represents the ratio 5 to 7.
   - **(b)** How many equal-sized parts are in the tape diagram in part (a)?
   - **(c)** Each equal-sized part of the tape diagram in part (a) represents how many people?
   - **(d)** How many children are at the park?
3. The ratio of the number of horses to the number of cows on a farm is 2 to 5. If there are 6 horses on the farm, use a ratio table to determine the number of cows on the farm.
4. The ratio of the number of boys to the number of girls in the gym is 8 to 12. If there are 36 girls in the gym, use a ratio table to determine the number of boys in the gym.

5. Tiana picked 20 flowers from her garden. She picked pink and yellow flowers. The ratio of the number of pink flowers to the number of yellow flowers is 3 to 2. How many pink flowers did Tiana pick?
   (a) Use a tape diagram to answer the question.
   (b) Use a ratio table to answer the question.

Solutions

1. **(a)**

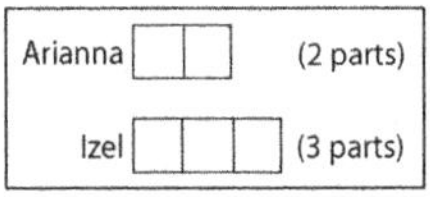

   (b) Arianna has 8 apples, so each equal-sized part of the tape diagram in part (a) represents $8 \div 2 = 4$ apples.
   (c) Arianna has 8 apples. Izel has $3 \times 4 = 12$ apples as shown below.

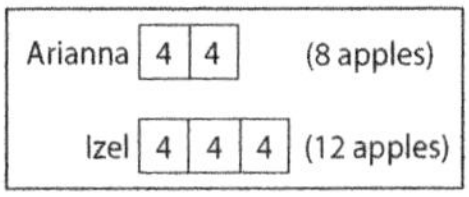

2. **(a)**

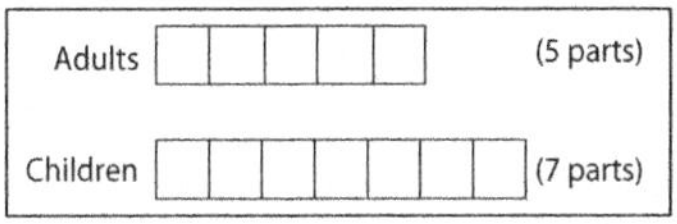

   (b) There are 12 equal-sized parts in the tape diagram.
   (c) The total number of people at the park is 108. Each equal-sized part of the tape diagram represents $108 \div 12 = 9$ people.
   (d) There are $7 \times 9 = 63$ children at the park as shown below.

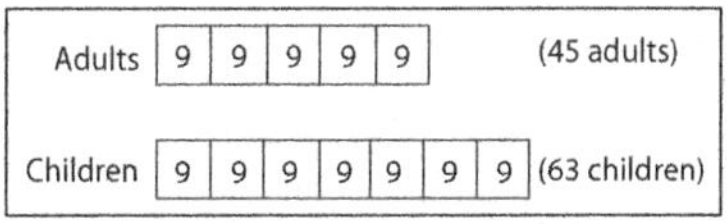

3. Create a ratio table. The table shows that for 6 horses, there are 15 cows.

Number of Horses	Number of Cows
2	5
4	10
6	15

4. Create a ratio table. The table shows that for every 36 girls, there are 24 boys.

Number of Boys	Number of Girls
8	12
16	24
24	36

5. **(a)** The tape diagram shows Tiana picked 12 pink flowers.

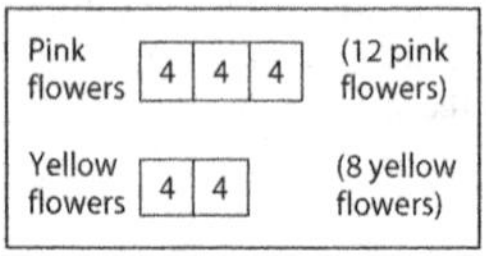

(b) The ratio table shows Tiana picked 12 pink flowers.

Number of Pink Flowers	Number of Yellow Flowers	Total Number of Flowers
3	2	5
6	4	10
9	6	15
12	8	20

Simplifying Ratios

(CCSS.Math.Content.6.RP.A.1, CCSS.Math.Content.6.RP.A.3.A, CCSS.Math.Content.6.RP.A.3.D)

You can write a ratio as a fraction. To simplify the ratio, simplify the fraction. You simplify fractions by dividing the numerator and denominator of the fraction by the largest number that will divide evenly into both. The result is an equivalent ratio in simplest form. There are 15 cows on the farm. There are 66 total animals on the farm. As a fraction, the ratio of the number of cows to the number of total animals on the farm is $\frac{15}{66}$. In simplest form, this ratio is $\frac{15 \div 3}{66 \div 3} = \frac{5}{22}$. Thus, you can say, for every 5 cows, there are 22 animals. Or you might say, for every 22 animals, 5 of them are cows.

When you find the ratio of two quantities, you must make sure they have the same units. When you write the ratio, the units will divide out. For example, the ratio of 2 pints to 5 quarts is *not* $\frac{2}{5}$ because these quantities are not expressed in the same units. From the measurement unit conversion table in the Appendix, 1 qt = 2 pt. Therefore, the ratio of 2 pt to 5 qt equals the ratio of 1 qt to 5 qt. This ratio is $\frac{1\text{ qt}}{5\text{ qt}} = \frac{1\,\cancel{\text{qt}}}{5\,\cancel{\text{qt}}} = \frac{1}{5}$ or 1 to 5 or 1:5.

☞ Try These

1. Simplify the given ratio. Write your answer as a fraction.
   - **(a)** the ratio of 25 to 35
   - **(b)** the ratio of 10 to 6
   - **(c)** the ratio of 6 to 16

2. Write ratios for the following descriptions. Write your answers in simplest fractional form.
   - **(a)** the ratio of 1 gallon to 12 quarts
   - **(b)** the ratio of 1 foot to 8 inches

Solutions

1. **(a)** Write the ratio as the fraction: $\frac{25}{35}$. Simplify: $\frac{25 \div 5}{35 \div 5} = \frac{5}{7}$.

   (b) Write the ratio as the fraction: $\frac{10}{6}$. Simplify: $\frac{10 \div 2}{6 \div 2} = \frac{5}{3}$.

   (c) Write the ratio as the fraction: $\frac{6}{16}$. Simplify: $\frac{6 \div 2}{16 \div 2} = \frac{3}{8}$.

2. **(a)** Gallons and quarts are not the same units. From the table in the Appendix, 1 gal = 4 qt. Thus, the ratio 1 gallon to 12 quarts equals the ratio of 4 qt to 12 qt. This ratio is $\frac{4\text{ qt}}{12\text{ qt}} = \frac{4\,\cancel{\text{qt}}}{12\,\cancel{\text{qt}}} = \frac{4}{12}$. In simplest form, it is $\frac{4 \div 4}{12 \div 4} = \frac{1}{3}$.

   (b) Feet and inches are not the same units. From the table in the Appendix, 1 ft = 12 in. Thus, the ratio of 1 foot to 8 inches equals the ratio of 12 in to 8 in. This ratio is $\frac{12\text{ in}}{8\text{ in}} = \frac{12\,\cancel{\text{in}}}{8\,\cancel{\text{in}}} = \frac{12}{8}$. In simplest form, it is $\frac{12 \div 4}{8 \div 4} = \frac{3}{2}$.

Using the Value of a Ratio

(CCSS.Math.Content.6.RP.A.1, CCSS.Math.Content.6.RP.A.3.A)

A ratio's **value** is the simplest form of the fraction that represents the ratio. In general, the value $\frac{a}{b}$ of a ratio of quantities A to B, in which A is greater than B, tells you A is $\frac{a}{b}$ times larger than B. If A is less than B, the value tells you A is the fraction $\frac{a}{b}$ of B. Here are examples.

> The value of the ratio of Carly's salary to Liana's salary is $\frac{2}{1}$. So Carly's salary is two times Liana's salary.
>
> The value of the ratio of Carly's height to Liana's height is $\frac{4}{5}$. So Carly is $\frac{4}{5}$ as tall as Liana.

Ratios that have the same value are equivalent ratios. As a fraction, the ratio 6 to 8 is $\frac{6}{8}$. The value of this ratio is $\frac{6 \div 2}{8 \div 2} = \frac{3}{4}$. As a fraction, the ratio 18 to 24 is $\frac{18}{24}$. The value of this ratio is $\frac{18 \div 6}{24 \div 6} = \frac{3}{4}$. The ratios 6 to 8 and 18 to 24 have the same value. They are equivalent ratios.

A recipe calls for 6 tablespoons of soy sauce and 18 tablespoons of brown sugar. The value of this ratio is $\frac{6 \div 6}{18 \div 6} = \frac{1}{3}$. Using this value, you can say the amount of soy sauce needed is $\frac{1}{3}$ times the amount of brown sugar needed. Or you can say the amount of brown sugar needed is 3 times the amount of soy sauce needed.

☞ Try These

1. Use the ratio's value to decide whether the ratio is equivalent to the ratio 8 to 18.
   (a) 16 to 54
   (b) 20 to 40
   (c) 40 to 90
   (d) 4 to 9

2. A biscuit recipe requires 3 cups of flour for 18 biscuits. Fill in the blank for each of the following.
   (a) The value of the ratio of the number of cups of flour to the number of biscuits is ________.
   (b) The number of cups of flour needed is ________ times the number of biscuits produced.
   (c) The number of biscuits produced is ________ times the number of cups of flour needed.
   (d) Making 54 biscuits would require ________ cups of flour.

3. The following ratio table shows the relationship between the number of pairs of a certain style of athletic shoe and the cost in dollars (not including sales tax). Fill in the missing elements.

Number of Pairs	Cost in Dollars
2	220
6	**(a)** ?
(b) ?	550

4. Which ratio has the greater value: A ratio of 120 to 25 or a ratio of 150 to 30?

5. Katy makes a sugar-water mixture using 14 cups of sugar and 28 cups of water. Ace makes a sugar-water mixture using 15 cups of sugar and 45 cups of water. Which mixture will taste sweeter?

Solutions

1. The value of the ratio 8 to 18 is $\frac{8 \div 2}{18 \div 2} = \frac{4}{9}$.

   (a) The value of the ratio 16 to 54 is $\frac{16 \div 2}{54 \div 2} = \frac{8}{27}$. Not equivalent.

   (b) The value of the ratio 20 to 40 is $\frac{20 \div 20}{40 \div 20} = \frac{1}{2}$. Not equivalent.

   (c) The value of the ratio 40 to 90 is $\frac{40 \div 10}{90 \div 10} = \frac{4}{9}$. Equivalent.

   (d) The value of the ratio 4 to 9 is $\frac{4}{9}$. Equivalent.

2. **(a)** $\frac{3 \div 3}{18 \div 3} = \frac{1}{6}$

   (b) $\frac{1}{6}$

   (c) 6

   (d) The number of cups of flour needed is $\frac{1}{6}$ times the number of biscuits produced. Therefore, making 54 biscuits would require $\frac{1}{6} \times 54 = \frac{1}{\cancel{6}_1} \times \frac{\cancel{54}^9}{1} = \frac{9}{1} = 9$ cups of flour.

3. In the first row, the ratio 2 to 220 has value $\frac{2 \div 2}{220 \div 2} = \frac{1}{110}$. Therefore, the number of pairs is $\frac{1}{110}$ times the cost. The cost in dollars is 110 times the number of pairs.

Number of Pairs	Cost in Dollars
2	220
(b) $\frac{1}{110} \times 550 = \frac{1}{110} \times \frac{550}{1} = \frac{550 \div 110}{110 \div 110} = \frac{5}{1} = 5$	550
6	**(a)** $6 \times 110 = 660$

4. The value of the ratio 120 to 25 is $\frac{120 \div 5}{25 \div 5} = \frac{24}{5}$. The value of the ratio 150 to 30 is $\frac{150 \div 30}{30 \div 30} = \frac{5}{1}$. The value $\frac{5}{1}$ is greater than the value $\frac{24}{5}$ because $\frac{5}{1} = \frac{5 \times 5}{1 \times 5} = \frac{25}{5}$.

5. The value of the ratio of sugar to water in Katy's mixture is $\frac{14 \div 14}{28 \div 14} = \frac{1}{2}$. The value of the ratio of sugar to water in Ace's mixture is $\frac{15 \div 15}{45 \div 15} = \frac{1}{3}$. The amount of sugar in Katy's mixture is $\frac{1}{2}$ times the amount of water. The amount of sugar in Ace's mixture is only $\frac{1}{3}$ times the amount of water. Therefore, Katy's mixture has more sugar for every cup of water, so it will taste sweeter. ***Tip:*** Recall, $\frac{1}{2} = \frac{3}{6}$ and $\frac{1}{3} = \frac{2}{6}$. So, $\frac{1}{2} > \frac{1}{3}$ because $\frac{3}{6} > \frac{2}{6}$.

Writing Equations for Ratios

(CCSS.Math.Content.6.RP.A.1, CCSS.Math.Content.6.RP.A.3.A, CCSS.Math.Content.6.RP.A.3.D)

If you know the value of a ratio, you can quickly write two equations to represent the ratio. In a recipe, the ratio of lemon juice to sugar is 3 to 1. The value of this ratio is $\frac{3}{1}$. The value means the amount of lemon juice (L) is 3 times the amount of sugar (S). Therefore, $L = 3S$. (***Tip:*** You do not need to put a times sign between 3 and S on the right side of the equation. When you write $3S$, it means 3 times S.) The value also means the amount of sugar (S) is $\frac{1}{3}$ times the amount of lemon juice (L). So, $S = \frac{1}{3}L$. The two equations $L = 3S$ and $S = \frac{1}{3}L$ are equivalent equations representing the same ratio.

The ratio of the number of apples to the number of pears is 6 to 8. The value of the ratio is $\frac{6 \div 2}{8 \div 2} = \frac{3}{4}$. The number of apples ($A$) is $\frac{3}{4}$ times the number of pears (P). The number of pears (P) is $\frac{4}{3}$ times the number of apples (A). The two equivalent equations are $A = \frac{3}{4}P$ and $P = \frac{4}{3}A$.

You might have to figure out the equation for a ratio by examining a ratio table. Here are examples.

Astrid is mixing paint for a remodeling project. The ratio table below shows the amount of green paint compared to the amount of blue paint. The first column is labeled "G" for the amount of green paint and the second column is labeled "B" for the amount of blue paint. The third column labeled "Relationship" describes the relationship between the B column entries and the G column entries.

Astrid has 8 pints of green paint. Determine the amount of blue paint she will need for her paint mixture.

G	B	Relationship
1	4	$4 = 4 \times 1$
2	8	$8 = 4 \times 2$
3	12	$12 = 4 \times 3$
4	16	$16 = 4 \times 4$
5	20	$20 = 4 \times 5$

In the table, the amount of blue paint is always 4 times the amount of green paint. Using variable names, you express this relationship as B is 4 times G. How would you write this relationship as an equation? The equation is $B = 4G$.

List the entries in the table as ordered pairs: (1, 4), (2, 8), (3, 12), (4, 16), (5, 20).

Plot the relationship $B = 4G$ in a coordinate plane. Connect the points because Astrid can use a portion of a quantity of paint (such as one-half of a container).

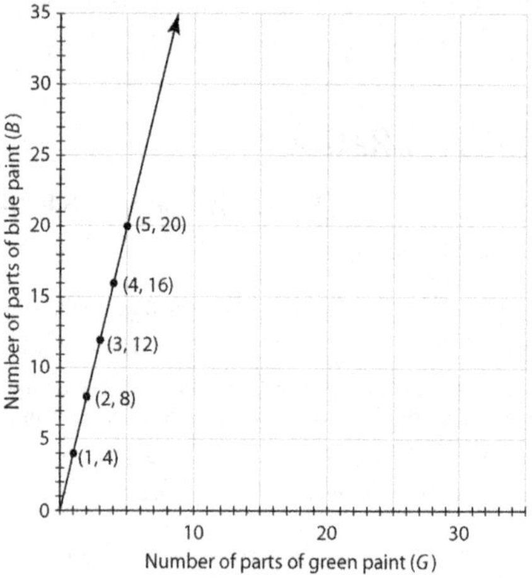

Astrid has 8 pints of green paint. Use the equation $B = 4G$ to determine the amount of blue paint she will need for her paint mixture: $B = 4 \times 8 = 32$. Astrid will need 32 pints of blue paint. You can see on the graph that (8, 32) would lie on the line representing $B = 4G$.

The number of cows compared to the number of horses on a farm is shown in the ratio table below. The first column is labeled "C" for the number of cows and the second column is labeled "H" for the number of horses. The third column labeled "Relationship" describes the relationship between the H column entries and the C column entries.

If there are 30 cows on the farm, how many horses are there?

C	*H*	**Relationship**
5	2	$2=\frac{2}{5}\times 5\left(\text{because } \frac{2}{{}_1\cancel{5}}\times\frac{\cancel{5}^1}{1}=\frac{2}{1}=2\right)$
10	4	$4=\frac{2}{5}\times 10\left(\text{because } \frac{2}{{}_1\cancel{5}}\times\frac{\cancel{10}^2}{1}=\frac{4}{1}=4\right)$
15	6	$6=\frac{2}{5}\times 15\left(\text{because } \frac{2}{{}_1\cancel{5}}\times\frac{\cancel{15}^3}{1}=\frac{6}{1}=6\right)$
20	8	$8=\frac{2}{5}\times 20\left(\text{because } \frac{2}{{}_1\cancel{5}}\times\frac{\cancel{20}^4}{1}=\frac{8}{1}=8\right)$
25	10	$10=\frac{2}{5}\times 25\left(\text{because } \frac{2}{{}_1\cancel{5}}\times\frac{\cancel{25}^5}{1}=\frac{10}{1}=10\right)$

In the table, the number of horses is always $\frac{2}{5}$ times the number of cows. Using variable names, you express this relationship as H is $\frac{2}{5}$ times C. The equation is $H=\frac{2}{5}C$. List the entries in the table as ordered pairs: (5, 2), (10, 4), (15, 6), (20, 8), (25, 10). Plot the relationship $H=\frac{2}{5}C$ in a coordinate plane. Do not connect the points because a portion of a cow or horse is unrealistic.

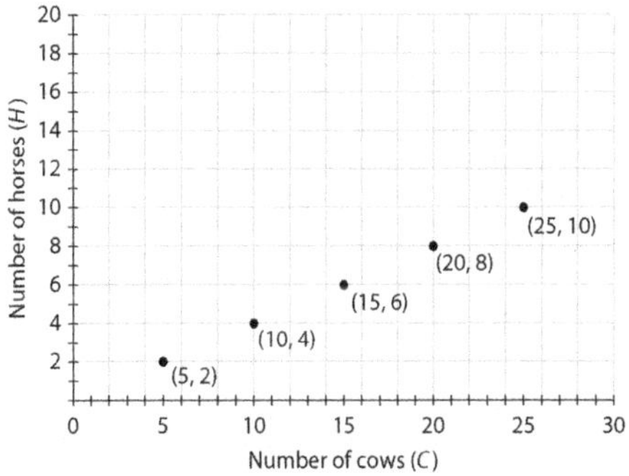

Use the equation to determine the number of horses corresponding to 30 cows:
$H=\frac{2}{5}\times 30=\frac{2}{{}_1\cancel{5}}\times\frac{\cancel{30}^6}{1}=\frac{12}{1}=12$. The number of horses corresponding to 30 cows is 12. You can see on the graph (30, 12) would line up with the other plotted points that represent the relationship $H=\frac{2}{5}C$.

☞ Try These

1. The ratio of the number of girls (G) to the number of boys (B) in a club is 7 to 3. Fill in the blank for each of the following.
   (a) The number of girls (G) is always __________ times the number of boys (B).
   (b) An equation that expresses the relationship in part (a) is $G =$ __________.
   (c) The number of boys (B) is always __________ times the number of girls (G).
   (d) An equation that expresses the relationship in part (c) is $B =$ __________.
   (e) If the number of boys in the club is 18, the number of girls in the club is __________.

2. Gabriel and Nahla work together raking and bagging leaves to earn money when they are not in school.
   (a) The following table shows Gabriel and Nahla's earnings based on the number of hours worked. Complete the table.

Number of Hours Worked	Amount Earned in Dollars	Relationship
2	16	
4	32	
5	40	
7	56	

   (b) Write an equation that will allow you to calculate Gabriel and Nahla's earnings in dollars (E) for any number of hours worked (H).
   (c) If Gabriel and Nahla work 6 hours as a team, how much money will they earn?

3. A party punch recipe uses 2 cups lime sherbet (S) for every 3 cups of ginger ale (G).
   (a) Complete the following table, where S is the amount of lime sherbet in cups and G is the amount of ginger ale in cups.

S	G	Relationship	Ordered Pairs
2	3		
4	6		
6	9		
8	12		
10	15		

   (b) Graph the ordered pairs.
   (c) Write an equation from which you can determine the amount of ginger ale (G) for any amount of lime sherbet (S).
   (d) How many cups of ginger ale are required for 16 cups of lime sherbet?

Solutions

1. **(a)** $\frac{7}{3}$

   (b) $G = \frac{7}{3}B$

   (c) $\frac{3}{7}$

   (d) $B = \frac{3}{7}G$

   (e) Use the equation $G = \frac{7}{3}B$ to answer the question. $G = \frac{7}{3}B = \frac{7}{3} \times 18 = \frac{7}{\cancel{3}_1} \times \frac{\cancel{18}^6}{1} = \frac{42}{1} = 42$

2. **(a)**

Number of Hours Worked	Amount Earned in Dollars	Relationship
2	16	$16 = 8 \times 2$
4	32	$32 = 8 \times 4$
5	40	$40 = 8 \times 5$
7	56	$56 = 8 \times 7$

(b) The table shows the amount earned (E) is always 8 times the number of hours worked (H). The equation is $E = 8H$.

(c) $E = 8H = 8 \times 6 = 48$. If Gabriel and Nahla work 6 hours, they will earn \$48.

3. **(a)**

S	G	Relationship	Ordered Pairs
2	3	$3 = \frac{3}{2} \times 2 \left(\text{because } \frac{3}{_1\cancel{2}} \times \frac{\cancel{2}^1}{1} = \frac{3}{1} = 3 \right)$	(2, 3)
4	6	$6 = \frac{3}{2} \times 4 \left(\text{because } \frac{3}{_1\cancel{2}} \times \frac{\cancel{4}^2}{1} = \frac{6}{1} = 6 \right)$	(4, 6)
6	9	$9 = \frac{3}{2} \times 6 \left(\text{because } \frac{3}{_1\cancel{2}} \times \frac{\cancel{6}^3}{1} = \frac{9}{1} = 9 \right)$	(6, 9)
8	12	$12 = \frac{3}{2} \times 8 \left(\text{because } \frac{3}{_1\cancel{2}} \times \frac{\cancel{8}^4}{1} = \frac{12}{1} = 12 \right)$	(8, 12)
10	15	$3 = \frac{3}{2} \times 10 \left(\text{because } \frac{3}{_1\cancel{2}} \times \frac{\cancel{10}^5}{1} = \frac{15}{1} = 15 \right)$	(10, 15)

(b)

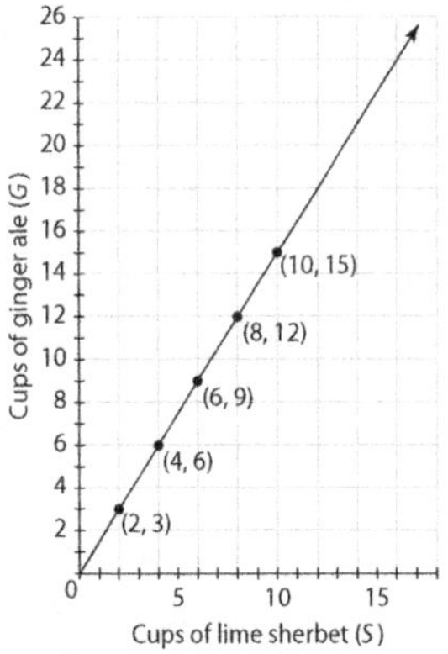

(c) $G = \frac{3}{2}S$

(d) $G = \frac{3}{2}S = \frac{3}{2} \times 16\text{ c} = \frac{3}{_1\cancel{2}} \times \frac{\cancel{16}^8\text{ c}}{1} = \frac{24\text{ c}}{1} = 24\text{ c}$. Thus, 24 cups of ginger ale are required for 16 cups of lime sherbet.

Understanding the Concept of a Rate

(CCSS.Math.Content.6.RP.A.2)

A **rate** is a multiplicative comparison of two quantities that have different measurement units. Suppose a car travels 300 miles in 5 hours. The **rate** is $\frac{300 \text{ miles}}{5 \text{ hours}}$. *Miles* and *hours* are different measurement units. They do not cancel out. The rate per 1 hour is $\frac{(300 \div 5) \text{ miles}}{(5 \div 5) \text{ hours}} = \frac{60 \text{ miles}}{1 \text{ hour}}$. This rate tells you the number of miles traveled in 1 hour. You can write "60 miles per hour" as $\frac{60 \text{ miles}}{1 \text{ hour}}$. The numerical factor of this quantity is $\frac{60}{1}$. In a **unit rate,** the second term (the denominator) of the numerical part is always 1. The unit rate tells the number of units of the first quantity for each 1 unit of the second quantity. In this example, there are 60 miles of distance for each 1 hour of time. The measurement unit for the unit rate is $\frac{\text{miles}}{\text{hour}}$. The rate $\frac{60 \text{ miles}}{1 \text{ hour}}$ is 60 $\frac{\text{miles}}{\text{hour}}$ or 60 miles per hour.

A **double-number-line diagram** allows you to illustrate a relationship in which the quantities have different units (for example, *miles* and *hours*). You can use a double-number-line diagram to show the relationship between miles and hours for a car traveling at a constant rate of 60 $\frac{\text{miles}}{\text{hour}}$. Mark equally spaced intervals on both number lines. Because the rate per 1 hour is 60 $\frac{\text{miles}}{\text{hour}}$, use miles as the units of the upper number line and hours as the units of the lower number line. Starting at zero, scale the upper number line in multiples of 60 miles because the rate is 60 miles per one time interval. Starting at zero, scale the lower number line in multiples of 1 hour because the time interval for the unit rate is 1 hour.

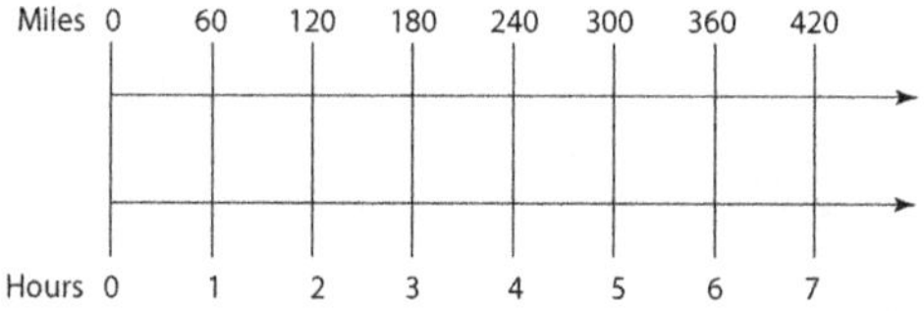

As you can see from the diagram, 0 hours corresponds to 0 miles, 1 hour corresponds to 60 miles, 2 hours corresponds to 120 miles, and so on.

Another way to illustrate the relationship that has a rate of 60 $\frac{\text{miles}}{\text{hour}}$ is to plot the points (0, 0), (1, 60), (2, 120), and so on in a coordinate plane. Use hours as the units of the horizontal axis and miles as the units of the vertical axis. Label the horizontal axis "Time in hours" and the vertical axis "Distance in miles."

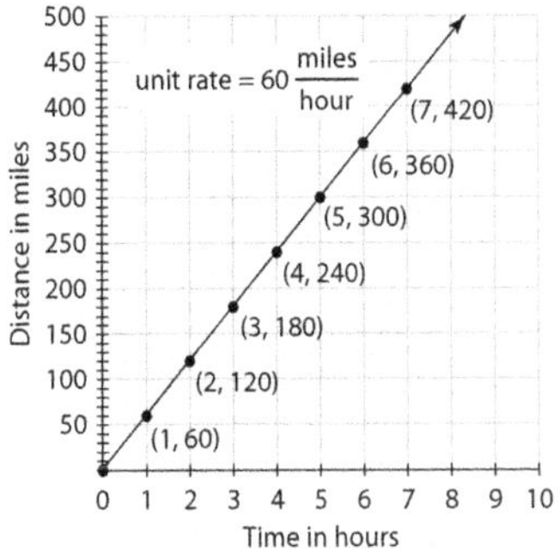

☞ Try These

1. Oksana runs 900 meters in 6 minutes. What is Oksana's rate for 1 minute?
2. Oliver mows 3 lawns in 2 hours. What is Oliver's lawn-mowing rate per hour?
3. Marta and Arnie, working as a team, clean 24 swimming pools in 4 days. What is their pool-cleaning rate for 1 day?
4. Jai paid \$74.85 for 15 frozen pepperoni pizzas. What was the cost per pizza?
5. Lucas can type 210 words in 3 minutes.
   (a) How many words per minute can Lucas type?
   (b) Using a double-number-line diagram, illustrate the relationship between words and minutes.
   (c) In a coordinate plane, illustrate the relationship between words and minutes.

Solutions

1. $\frac{900 \text{ meters}}{6 \text{ minutes}} = \frac{(900 \div 6) \text{ meters}}{(6 \div 6) \text{ minutes}} = \frac{150 \text{ meters}}{1 \text{ minute}} = 150 \frac{\text{meters}}{\text{minute}} = 150 \text{ meters per minute}$
2. $\frac{3 \text{ lawns}}{2 \text{ hours}} = \frac{(3 \div 2) \text{ lawns}}{(2 \div 2) \text{ hours}} = \frac{1.5 \text{ lawns}}{1 \text{ hour}} = 1.5 \frac{\text{lawns}}{\text{hour}} = 1.5 \text{ lawns per hour}$
3. $\frac{24 \text{ pools}}{4 \text{ days}} = \frac{(24 \div 4) \text{ pools}}{(4 \div 4) \text{ days}} = \frac{6 \text{ pools}}{1 \text{ day}} = 6 \frac{\text{pools}}{\text{day}} = 6 \text{ pools per day}$
4. $\frac{\$74.85}{15 \text{ pizzas}} = \frac{\$74.85 \div 15}{(15 \div 15) \text{ pizzas}} = \frac{\$4.99}{1 \text{ pizza}} = \$4.99 \text{ per pizza}$

5. **(a)** $\frac{210 \text{ words}}{3 \text{ minutes}} = \frac{(210 \div 3) \text{ words}}{(3 \div 3) \text{ minutes}} = \frac{70 \text{ words}}{1 \text{ minute}} = 70 \frac{\text{words}}{\text{minute}} = 70$ words per minute

(b)

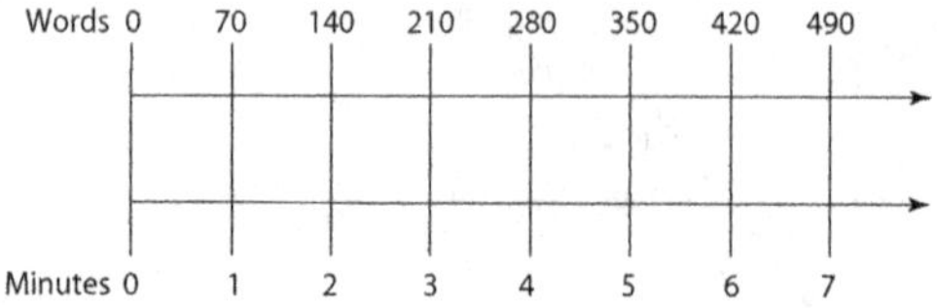

(c)

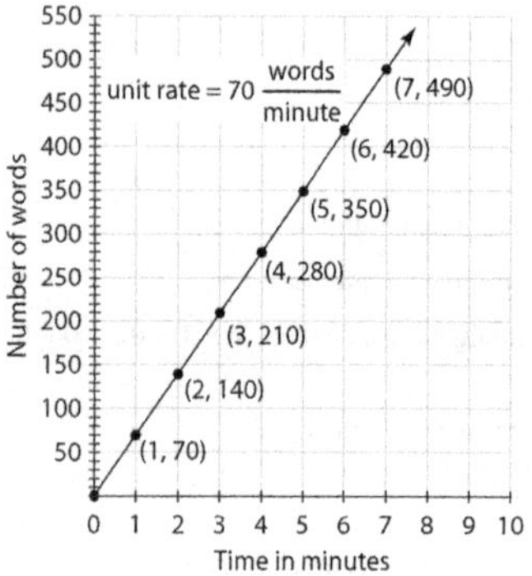

Using Unit Price for Comparison Shopping

(CCSS.Math.Content.6.RP.A.2, CCSS.Math.Content.6.RP.A.3.B)

The **unit price** is the price for one unit of an item. For party favors, Ansel buys 12 miniature toy cars at Store X for \$11.64. The unit price at Store X for the toy cars is $\frac{\$11.64}{12 \text{ toy cars}} = \frac{\$11.64 \div 12}{(12 \div 12) \text{ toy cars}} = \frac{\$0.97}{1 \text{ toy car}} = \$0.97$ per toy car.

In **comparison shopping** problems, you compare the unit price of similar items. The better deal is the one that has the lower unit price. Of course, you naturally assume there is no difference in the quality of the items whose prices you are comparing.

Suppose Ansel's friend Juno tells him he can buy 13 of the same miniature toy cars at Store Y for \$12.74. Which store has the better deal for the miniature toy cars? To answer the question, compute the unit price at Store Y for the toy cars. The unit price is $\frac{\$12.74}{13 \text{ toy cars}} = \frac{\$12.74 \div 13}{(13 \div 13) \text{ toy cars}} = \frac{\$0.98}{1 \text{ toy car}} = \$0.98$ per toy car.

The unit price is higher at Store Y, so Store X has the better deal.

☞ Try These

1. Which is the better deal for disposable ballpoint pens: Five for \$6.25 or six for \$7.38?
2. Which is the better deal for graphic tees: Two for \$64 or three for \$97.50?
3. Which is the better deal for fresh strawberries: Three pounds for \$7.98 or four pounds for \$11.96?
4. Which is the better deal for athletic socks: Five pairs for \$44.90 or six pairs for \$53.94?
5. Which is the better deal for sunscreen: Five fluid ounces for \$7.05 or twelve fluid ounces for \$16.80?

Solutions

1. $\dfrac{\$6.25}{5 \text{ pens}} = \dfrac{\$6.25 \div 5}{(5 \div 5) \text{ pens}} = \dfrac{\$1.25}{1 \text{ pen}} = \$1.25 \text{ per pen}$

   $\dfrac{\$7.38}{6 \text{ pens}} = \dfrac{\$7.38 \div 6}{(6 \div 6) \text{ pens}} = \dfrac{\$1.23}{1 \text{ pen}} = \$1.23 \text{ per pen}$

   Six pens for \$7.38 is the better deal.

2. $\dfrac{\$64}{2 \text{ tees}} = \dfrac{\$64 \div 2}{(2 \div 2) \text{ tees}} = \dfrac{\$32}{1 \text{ tee}} = \$32 \text{ per tee}$

   $\dfrac{\$97.50}{3 \text{ tees}} = \dfrac{\$97.50 \div 3}{(3 \div 3) \text{ tees}} = \dfrac{\$32.50}{1 \text{ tee}} = \$32.50 \text{ per tee}$

   Two tees for \$64 is the better deal.

3. $\dfrac{\$7.98}{3 \text{ pounds}} = \dfrac{\$7.98 \div 3}{(3 \div 3) \text{ pounds}} = \dfrac{\$2.66}{1 \text{ pound}} = \$2.66 \text{ per pound}$

   $\dfrac{\$11.96}{4 \text{ pounds}} = \dfrac{\$11.96 \div 4}{(4 \div 4) \text{ pounds}} = \dfrac{\$2.99}{1 \text{ pound}} = \$2.99 \text{ per pound}$

   Three pounds for \$7.98 is the better deal.

4. $\dfrac{\$44.90}{5 \text{ pairs}} = \dfrac{\$44.90 \div 5}{(5 \div 5) \text{ pairs}} = \dfrac{\$8.98}{1 \text{ pair}} = \$8.98 \text{ per pair}$

   $\dfrac{\$53.94}{6 \text{ pairs}} = \dfrac{\$53.94 \div 6}{(6 \div 6) \text{ pairs}} = \dfrac{\$8.99}{1 \text{ pair}} = \$8.99 \text{ per pair}$

   Five pairs for \$44.90 is the better deal.

5. $\dfrac{\$7.05}{5 \text{ fluid ounces}} = \dfrac{\$7.05 \div 5}{(5 \div 5) \text{ fluid ounces}} = \dfrac{\$1.41}{1 \text{ fluid ounce}} = \$1.41 \text{ per fluid ounce}$

   $\dfrac{\$16.80}{12 \text{ fluid ounces}} = \dfrac{\$16.80 \div 12}{(12 \div 12) \text{ fluid ounces}} = \dfrac{\$1.40}{1 \text{ fluid ounce}} = \$1.40 \text{ per fluid ounce}$

   Twelve fluid ounces for \$16.80 is the better deal.

Converting Measurement Units

(CCSS.Math.Content.6.RP.A.2, CCSS.Math.Content.6.RP.A.3.D)

You can convert from one measurement unit to another by using an appropriate **conversion rate.** You make conversion rates by using the conversion facts given in a measurement unit conversion table (see the Appendix for an example of a measurement unit conversion table). For each conversion fact in the table, you can write *two* conversion rates. For example, the table shows the conversion fact, 1 yd = 3 ft. The two conversion rates for this fact are $\frac{1\text{ yd}}{3\text{ ft}}$ and $\frac{3\text{ ft}}{1\text{ yd}}$. Each of these conversion rates is equivalent to the number 1 because the numerator and denominator are different names for the same length. Therefore, if you multiply a quantity by either of these conversion rates, you will not change the value of the quantity.

To convert one measurement unit to another unit, multiply the quantity to be converted by the conversion rate whose *denominator's units are the same as the units of the quantity to be converted.* This strategy is called **dimensional analysis.** When you do the multiplication, the units you started out with will cancel out. You will be left with the new units. If this doesn't happen, then you used the wrong conversion rate. Try multiplying with the other conversion rate.

It is a good idea to assess your final answer to see whether it makes sense. When you are converting from *a larger unit to a smaller unit,* you should expect it will take *more* of the smaller units to equal the same amount. When you are converting from *a smaller unit to a larger unit,* you should expect it will take *less* of the larger units to equal the same amount.

Here is an example of converting from a larger unit to a smaller unit.

How many feet are in 5 yards?

From the table in the Appendix, the conversion rates are $\frac{1\text{ yd}}{3\text{ ft}}$ and $\frac{3\text{ ft}}{1\text{ yd}}$. Write 5 yards as a fraction with denominator 1 and let dimensional analysis tell you whether to multiply by $\frac{1\text{ yd}}{3\text{ ft}}$ or $\frac{3\text{ ft}}{1\text{ yd}}$. Multiply by $\frac{3\text{ ft}}{1\text{ yd}}$ because its denominator has the same units as 5 yards, the quantity to be converted.

$$5\text{ yards} = \frac{5\text{ yd}}{1} \times \frac{3\text{ ft}}{1\text{ yd}} = \frac{5\ \cancel{\text{yd}}}{1} \times \frac{3\text{ ft}}{1\ \cancel{\text{yd}}} = \frac{15\text{ ft}}{1} = 15\text{ feet}$$

As you can see, the "yd" units cancel out, leaving "ft" as the units for the answer.

Does this answer make sense? Yes. Feet are smaller than yards, so it should take more of them to equal the same length as 5 yards.

Use a double-number-line diagram to support this result.

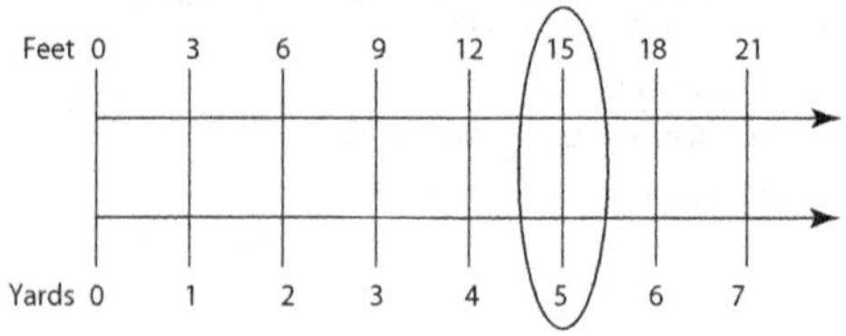

Here is an example of converting from a smaller unit to a larger unit.

How many meters is 200 centimeters?

From the table in the Appendix, the conversion rates are $\frac{1\text{ m}}{100\text{ cm}}$ and $\frac{100\text{ cm}}{1\text{ m}}$. Write 200 centimeters as a fraction with denominator 1 and let dimensional analysis tell you whether to multiply by $\frac{1\text{ m}}{100\text{ cm}}$ or $\frac{100\text{ cm}}{1\text{ m}}$. Multiply 200 centimeters by $\frac{1\text{ m}}{100\text{ cm}}$ because its denominator has the same units as 200 centimeters, the quantity to be converted.

$$200\text{ centimeters} = \frac{200\text{ cm}}{1} \times \frac{1\text{ m}}{100\text{ cm}} = \frac{200\ \cancel{\text{cm}}}{1} \times \frac{1\text{ m}}{100\ \cancel{\text{cm}}} = \frac{200\text{ m}}{100} = \frac{(200 \div 100)\text{ m}}{100 \div 100} = \frac{2\text{ m}}{1} = 2\text{ meters}$$

As you can see, the "cm" units cancel out, leaving "m" as the units for the answer.

Does this answer make sense? Yes. Meters are larger than centimeters, so it should take fewer of them to equal the same distance as 200 centimeters.

Use a double-number-line diagram to support this result.

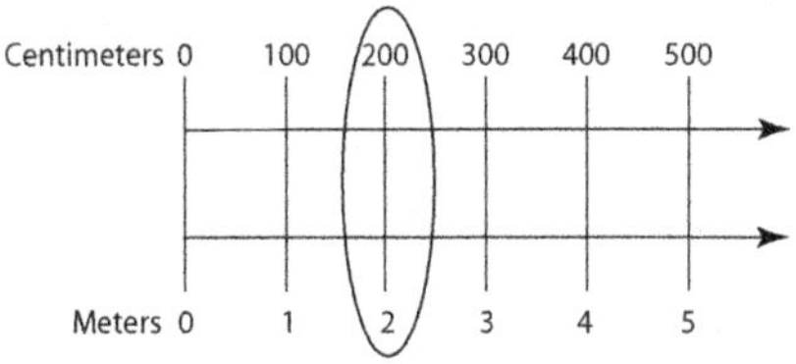

☞ Try These

1. **(a)** How many cups are in 3 quarts?
   (b) Support your answer to part (a) with a double-number-line diagram.
2. **(a)** How many grams are in 7 kilograms?
   (b) Support your answer to part (a) with a double-number-line diagram.
3. **(a)** How many feet are 48 inches?
   (b) Support your answer to part (a) with a double-number-line diagram.
4. **(a)** How many minutes are in 6 hours?
   (b) Support your answer to part (a) with a double-number-line diagram.
5. **(a)** How many kilometers are 1500 meters?
   (b) Support your answer to part (a) with a double-number-line diagram.

Solutions

1. **(a)** 3 quarts $= \frac{3 \text{ qt}}{1} \times \frac{4 \text{ c}}{1 \text{ qt}} = \frac{3 \cancel{\text{qt}}}{1} \times \frac{4 \text{ c}}{1 \cancel{\text{qt}}} = \frac{12 \text{ c}}{1} = 12$ cups

   (b)

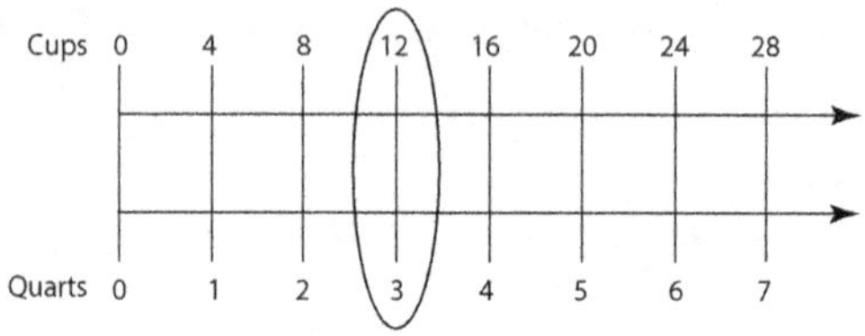

2. **(a)** 7 kilograms $= \frac{7 \text{ kg}}{1} \times \frac{1000 \text{ g}}{1 \text{ kg}} = \frac{7 \cancel{\text{kg}}}{1} \times \frac{1000 \text{ g}}{1 \cancel{\text{kg}}} = \frac{7000 \text{ g}}{1} = 7000$ grams

   (b)

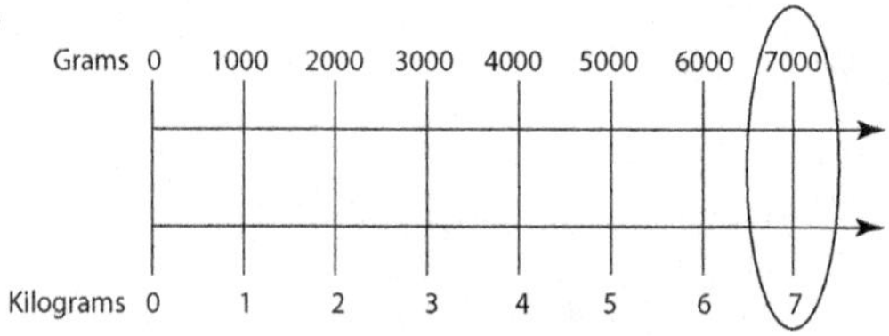

3. **(a)** 48 inches $= \frac{48 \text{ in}}{1} \times \frac{1 \text{ ft}}{12 \text{ in}} = \frac{48 \cancel{\text{in}}}{1} \times \frac{1 \text{ ft}}{12 \cancel{\text{in}}} = \frac{48 \text{ ft}}{12} = \frac{(48 \div 12) \text{ ft}}{12 \div 12} = \frac{4 \text{ ft}}{1} = 4$ feet

   (b)

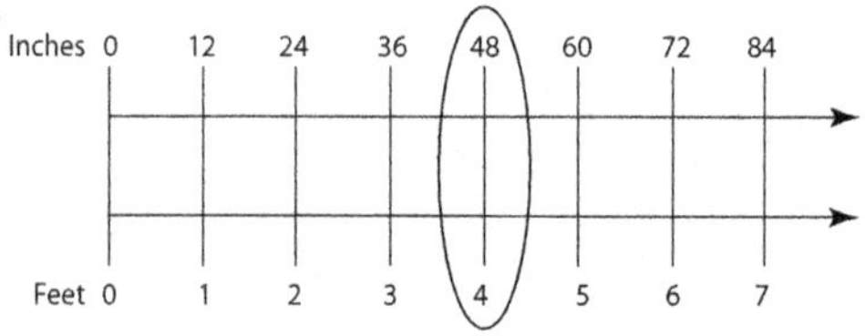

4. **(a)** 6 hours $= \frac{6 \text{ hr}}{1} \times \frac{60 \text{ min}}{1 \text{ hr}} = \frac{6 \cancel{\text{hr}}}{1} \times \frac{60 \text{ min}}{1 \cancel{\text{hr}}} = \frac{360 \text{ min}}{1} = 360$ minutes

   (b)

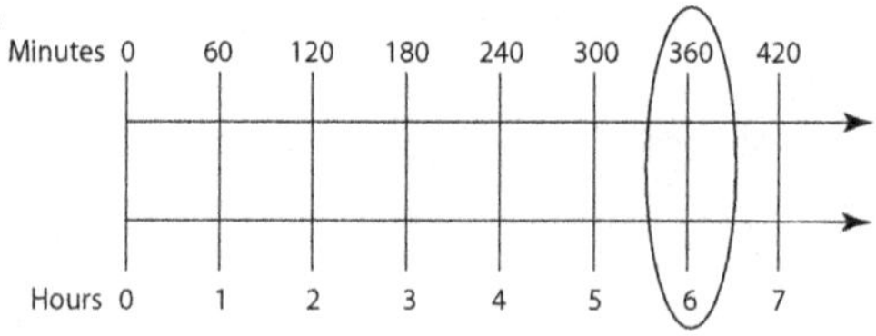

5. **(a)** $1500 \text{ meters} = \frac{1500 \text{ m}}{1} \times \frac{1 \text{ km}}{1000 \text{ m}} = \frac{1500 \cancel{\text{m}}}{1} \times \frac{1 \text{ km}}{1000 \cancel{\text{m}}} = \frac{1500 \text{ km}}{1000} = \frac{(1500 \div 1000) \text{ km}}{1000 \div 1000}$

$= \frac{1.5 \text{ km}}{1} = 1.5 \text{ kilometers}$

(b)

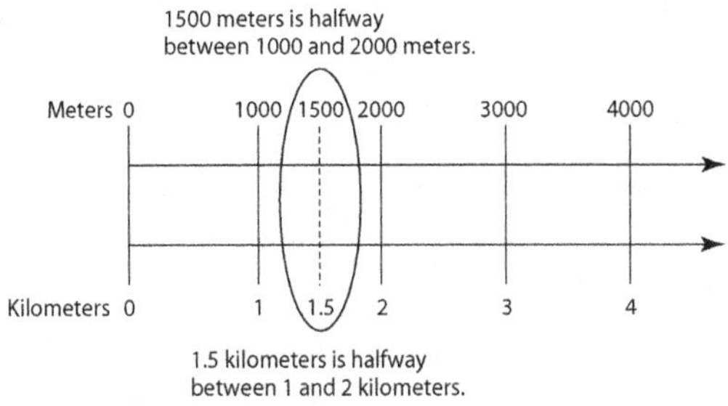

Solving Unit Rate Problems

(CCSS.Math.Content.6.RP.A.2, CCSS.Math.Content.6.RP.A.3.B, CCSS.Math.Content.6.RP.A.3.D)

Suppose Lucas types at the rate of $70\frac{\text{words}}{\text{minute}}$. Here is a question you might ask.

At this rate, how many words can Lucas type in 5 minutes?

Answer this question by multiplying the rate per minute by the amount of time in minutes Lucas types.

Tip: Write "70 words per minute" as $\frac{70 \text{ words}}{1 \text{ minute}}$ and write 5 minutes as $\frac{5 \text{ minutes}}{1}$. This strategy will help you deal with the units correctly.

$$70\frac{\text{words}}{\text{minute}} \times 5 \text{ minutes} = \frac{70 \text{ words}}{1 \cancel{\text{minute}}} \times \frac{5 \cancel{\text{minutes}}}{1} = \frac{350 \text{ words}}{1} = 350 \text{ words}$$

Lucas can type 350 words in 5 minutes.

Does the answer make sense? Yes. In the multiplication, the "minutes" units cancel out, leaving "words" as the units for the answer. So the units part of the answer makes sense. A double-number-line diagram also supports the answer.

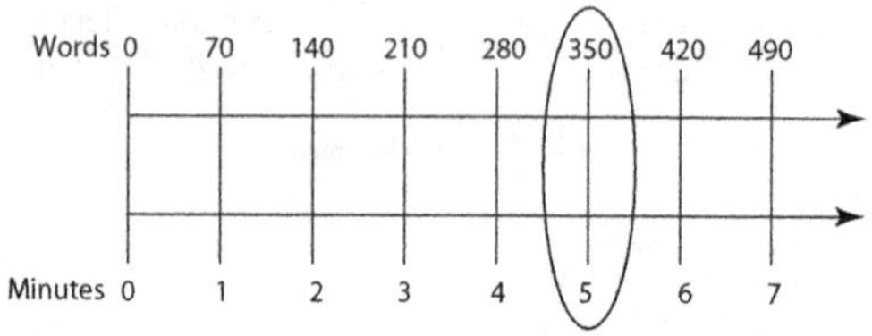

Here is a different question you could ask.

At this rate, how many minutes would it take Lucas to type 490 words?

To answer this question, write the rate, $\frac{70 \text{ words}}{1 \text{ minute}}$, as its equivalent rate of $\frac{1 \text{ minute}}{70 \text{ words}}$. Multiply 490 words by this equivalent rate. ***Tip:*** You use the equivalent rate so that the units will work out correctly.

$$490 \text{ words} \times \frac{1 \text{ minute}}{70 \text{ words}} = \frac{490 \cancel{\text{ words}}}{1} \times \frac{1 \text{ minute}}{70 \cancel{\text{ words}}} = \frac{490 \text{ minutes}}{70} = \frac{(490 \div 70) \text{ minutes}}{70 \div 70} = 7 \text{ minutes}$$

It will take Lucas 7 minutes to type 490 words.

Does this answer make sense? Yes. In the multiplication, the "words" units canceled out, leaving "minutes" as the units for the answer. So the units part of the answer makes sense. And, again, a double-number-line diagram supports the answer.

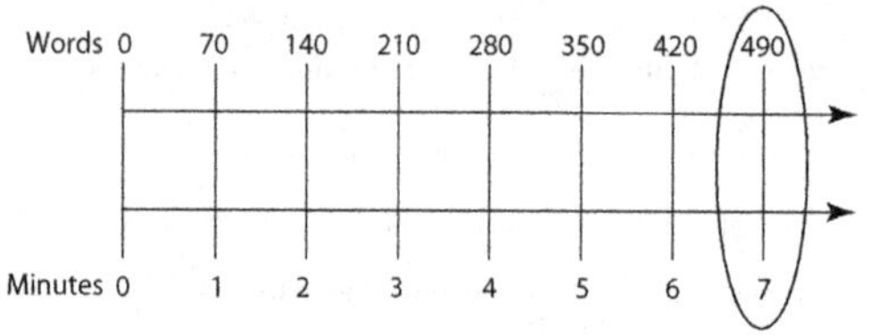

In some problems you are not given the rate per unit. Look at this example.

Suppose five folders cost \$7.50. At this rate, how much would seven folders cost?

Answering this question will take two steps. First, compute the rate for one folder. Next, multiply the rate by the number of folders.

Step 1. Compute the rate for one folder.

$$\frac{\$7.50}{5 \text{ folders}} = \frac{\$7.50 \div 5}{(5 \div 5) \text{ folders}} = \frac{\$1.50}{1 \text{ folder}}$$

Step 2. Multiply the rate by the number of folders.

$$\frac{\$1.50}{1 \text{ folder}} \times 7 \text{ folders} = \frac{\$1.50}{1 \cancel{\text{ folder}}} \times \frac{7 \cancel{\text{ folders}}}{1} = \frac{\$10.50}{1} = \$10.50$$

Seven folders will cost \$10.50.

☞ Try These

1. A car travels at the rate of $60\ \dfrac{\text{miles}}{\text{hour}}$.
   - **(a)** At this rate, how many miles will the car travel in 4 hours?
   - **(b)** Support your answer to part (a) with a double-number-line diagram.
   - **(c)** At this rate, how many hours will it take the car to travel 300 miles?
   - **(d)** Support your answer to part (c) with a double-number-line diagram.
2. Marta and Arnie, working as a team, clean 24 swimming pools in 4 days. At this rate, how many pools can Marta and Arnie clean in 2 days?
3. Jai paid $74.85 for 15 frozen pepperoni pizzas. At this rate, how much would 20 pizzas cost?
4. Oksana runs 900 meters in 6 minutes. At this rate, how many minutes will it take Oksana to run 600 meters?

Solutions

1. **(a)** $60\ \dfrac{\text{miles}}{\text{hour}} \times 4 \text{ hours} = \dfrac{60 \text{ miles}}{1\ \cancel{\text{hour}}} \times \dfrac{4\ \cancel{\text{hours}}}{1} = \dfrac{240 \text{ miles}}{1} = 240$ miles. The car will travel 240 miles in 4 hours.

   (b)

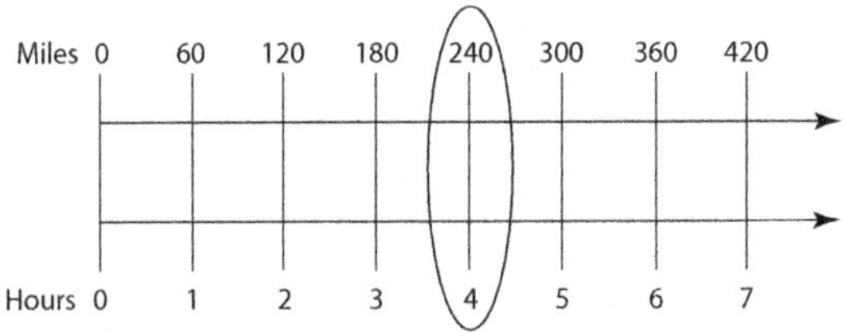

   (c) $300 \text{ miles} \times \dfrac{1 \text{ hour}}{60 \text{ miles}} = \dfrac{300\ \cancel{\text{miles}}}{1} \times \dfrac{1 \text{ hour}}{60\ \cancel{\text{miles}}} = \dfrac{300 \text{ hours}}{60} = \dfrac{(300 \div 60) \text{ hours}}{60 \div 60} = \dfrac{5 \text{ hours}}{1} = 5 \text{ hours}$

   (d) It will take 5 hours for the car to travel 300 miles.

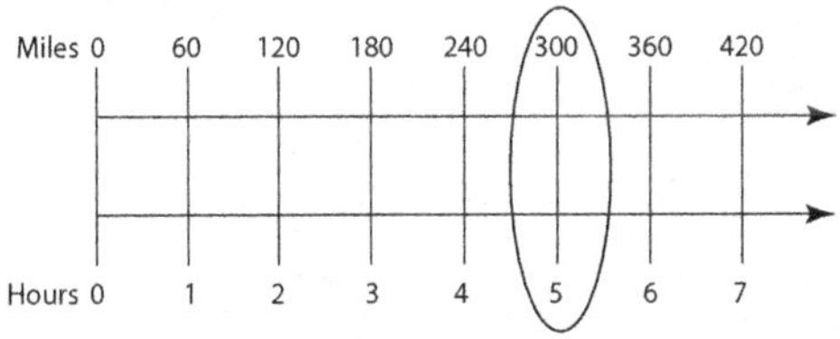

2. *Step 1.* Compute the rate per day.

$$\frac{24 \text{ pools}}{4 \text{ days}} = \frac{(24 \div 4) \text{ pools}}{(4 \div 4) \text{ days}} = \frac{6 \text{ pools}}{1 \text{ day}}$$

Step 2. Multiply the rate by the number of days.

$$\frac{6 \text{ pools}}{1 \text{ day}} \times 2 \text{ days} = \frac{6 \text{ pools}}{1 \cancel{\text{day}}} \times \frac{2 \cancel{\text{days}}}{1} = \frac{12 \text{ pools}}{1} = 12 \text{ pools}$$

Marta and Arnie can clean 12 pools in 2 days.

3. *Step 1.* Compute the price for one pizza.

$$\frac{\$74.85}{15 \text{ pizzas}} = \frac{\$74.85 \div 15}{(15 \div 15) \text{ pizzas}} = \frac{\$4.99}{1 \text{ pizza}} = \$4.99 \text{ per pizza}$$

Step 2. Multiply the price per pizza by the number of pizzas.

$$\frac{\$4.99}{1 \text{ pizza}} \times 20 \text{ pizzas} = \frac{\$4.99}{1 \cancel{\text{pizza}}} \times \frac{20 \cancel{\text{pizzas}}}{1} = \frac{\$99.80}{1} = \$99.80$$

Twenty pizzas will cost $99.80.

4. *Step 1.* Compute the rate for 1 minute.

$$\frac{900 \text{ meters}}{6 \text{ minutes}} = \frac{(900 \div 6) \text{ meters}}{(6 \div 6) \text{ minutes}} = \frac{150 \text{ meters}}{1 \text{ minute}}$$

Step 2. Multiply 600 meters by the equivalent rate of $\frac{1 \text{ minute}}{150 \text{ meters}}$.

$$600 \text{ meters} \times \frac{1 \text{ minute}}{150 \text{ meters}} = \frac{600 \cancel{\text{meters}}}{1} \times \frac{1 \text{ minute}}{150 \cancel{\text{meters}}} = \frac{600 \text{ minutes}}{150} = \frac{(600 \div 150) \text{ minutes}}{150 \div 150} =$$

$$\frac{4 \text{ minutes}}{1} = 4 \text{ minutes}$$

It will take Oksana 4 minutes to run 600 meters.

Understanding Percents

(CCSS.Math.Content.6.RP.A.1, CCSS.Math.Content.6.RP.A.3.C)

Percent means "per hundred" or "hundredths." A percent is a part-to-whole ratio in which the second element is 100. Suppose in a department store the ratio of men to the total number of people in the store is 30 to 100. There are 30 men per 100 people. So the ratio of men to the total number of people is $\frac{30}{100}$. You can express the ratio $\frac{30}{100}$ as 30%. You can say "the ratio of men to the total number of people is 30%" or "30% of the people in the store are men."

The expression 30% means $\frac{30}{100}$. A percent sign is a shorthand way to express hundredths. It means $\frac{1}{100}$. Thus, for example, $5\% = \frac{5}{100}$, $25\% = \frac{25}{100}$, $50\% = \frac{50}{100}$, and $100\% = \frac{100}{100}$ (or 1). So you see, percents are special fractions whose denominators are 100. The fractional unit is $\frac{1}{100}$.

On a number line, percents between 0% and 100% lie between 0 and 1. They are equivalent to fractions between 0 and 1. The following double-number-line diagram shows this relationship.

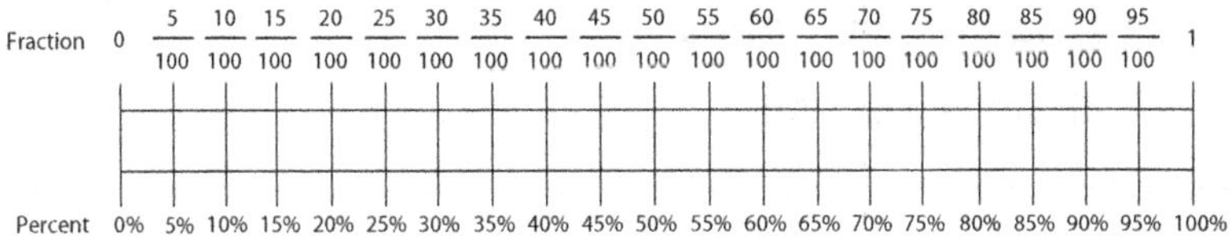

Try These

1. Write each percent as an equivalent fraction whose denominator is 100.
   (a) 1%
   (b) 10%
   (c) 45%
   (d) 90%

2. Fill in the blank with a fraction whose denominator is 100.
   (a) 25% is another way to express the fraction ________.
   (b) 30% is another way to express the fraction ________.
   (c) 50% is another way to express the fraction ________.
   (d) 75% is another way to express the fraction ________.

Solutions

1. **(a)** $1\% = \frac{1}{100}$

   (b) $10\% = \frac{10}{100}$

   (c) $45\% = \frac{45}{100}$

   (d) $90\% = \frac{90}{100}$

2. **(a)** $\frac{25}{100}$
   (b) $\frac{30}{100}$
   (c) $\frac{50}{100}$
   (d) $\frac{75}{100}$

Simplifying Percent Ratios

(CCSS.Math.Content.6.RP.A.1, CCSS.Math.Content.6.RP.A.3.C)

When you change a percent to a fraction whose denominator is 100, you often can simplify the fraction you get. For example, $5\% = \frac{5}{100} = \frac{5 \div 5}{100 \div 5} = \frac{1}{20}$, $25\% = \frac{25}{100} = \frac{25 \div 25}{100 \div 25} = \frac{1}{4}$, $60\% = \frac{60}{100} = \frac{60 \div 20}{100 \div 20} = \frac{3}{5}$, and $100\% = \frac{100}{100} = 1$.

Try These

1. Write each percent as an equivalent fraction in simplest form.
   (a) 10%
   (b) 45%
   (c) 50%
   (d) 90%

2. Fill in the blank with a fraction in simplest form.
   (a) 10% is another way to express the fraction __________.
   (b) 25% is another way to express the fraction __________.
   (c) 50% is another way to express the fraction __________.
   (d) 75% is another way to express the fraction __________.

Solutions

1. **(a)** $10\% = \frac{10}{100} = \frac{10 \div 10}{100 \div 10} = \frac{1}{10}$
   (b) $45\% = \frac{45}{100} = \frac{45 \div 5}{100 \div 5} = \frac{9}{20}$
   (c) $50\% = \frac{50}{100} = \frac{1}{2}$
   (d) $90\% = \frac{90}{100} = \frac{90 \div 10}{100 \div 10} = \frac{9}{10}$

2. **(a)** $\frac{1}{10}$

   (b) $\frac{1}{4}$

   (c) $\frac{1}{2}$

   (d) $\frac{3}{4}$

Solving Percent Problems

(CCSS.Math.Content.6.RP.A.1, CCSS.Math.Content.6.RP.A.3.C)

A percent relationship involves three elements: a percent, a part, and a whole. When you know two of these elements, you can solve for the third element. Use your understanding that the given percent equals $\frac{\text{part}}{\text{whole}}$. This relationship means the part is the given percent of the whole.

Tip: As you know, the percent is the quantity with the % sign attached. It cannot be used as such in a computation. Change it to an equivalent fraction before you do computations.

Finding the Part, Given the Percent and the Whole

A percent always represents a relationship between a part and a whole. If the percent is 30%, then the part is 30% of the whole. When you know the percent and the whole, you find the part by multiplying the percent times the whole. Of course, you must change the percent to an equivalent fraction before you do the actual computation. Look at this example.

> 30% of 200 people are men. How many of the people are men?

For this question, the percent is 30% and the whole is 200.

Tip: The whole is usually the amount that immediately follows the word "of."

The part is 30% times 200 $= \frac{30}{100} \times 200 = \frac{30}{\cancel{100}_1} \times \frac{\cancel{200}^2}{1} = \frac{60}{1} = 60$

There are 60 men.

Use a double-number-line diagram to support this result. Label the upper number line "Part" and the lower number line "Percent." The whole, 200, corresponds to 100%. Each 1% is $\frac{1}{100}$ of 200, which is 2. So each 10% is 10 times 2, which is 20. For convenience, mark the "Part" number line in increments of 20 and the "Percent" number line in corresponding increments of 10%.

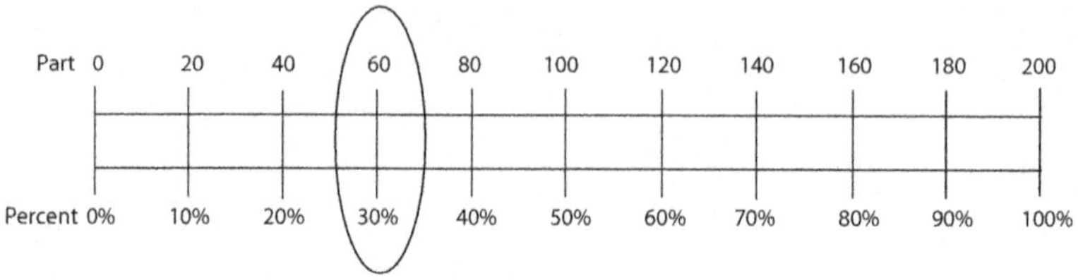

The diagram shows 60 is 30% of 200.

☞ Try These

1. **(a)** What is 50% of 180?
   (b) Support your answer to part (a) with a double-number-line diagram.
2. **(a)** In an auditorium of 500 students, 25% are sixth graders. How many of the students are sixth graders?
   (b) Support your answer to part (a) with a double-number-line diagram.
3. There are 80 vehicles in a parking lot. Fifteen percent of the vehicles are vans. How many vans are in the parking lot?
4. A digital video recorder is on sale for 10% off the original price. The original price is $399. What is the sale price of the recorder?

Solutions

1. **(a)** The percent is 50% and the whole is 180. The part is

$$50\% \text{ times } 180 = \frac{50}{100} \times 180 = \frac{1}{{}_1\cancel{2}} \times \frac{\cancel{180}^{90}}{1} = \frac{90}{1} = 90$$

   90 is 50% of 180.

   (b) Label the upper number line "Part" and the lower number line "Percent." The whole, 180, corresponds to 100%. Each 1% is $\frac{1}{100}$ of 180, which is 1.8. So each 10% is 10 times 1.8, which is 18. For convenience, mark the "Part" number line in increments of 18 and the "Percent" number line in corresponding increments of 10%.

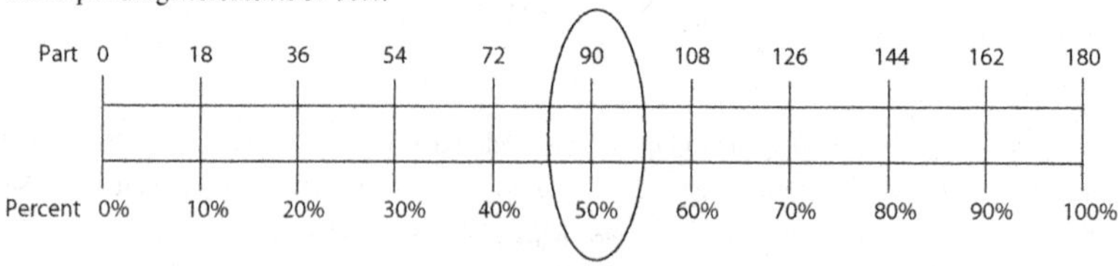

2. **(a)** The percent is 25% and the whole is 500. The part is

$$25\% \text{ times } 500 = \frac{25}{100} \times 500 = \frac{1}{{}_1\cancel{4}} \times \frac{\cancel{500}^{125}}{1} = \frac{125}{1} = 125$$

There are 125 sixth graders in the auditorium.

(b) Label the upper number line "Part" and the lower number line "Percent." The whole, 500, corresponds to 100%. Each 1% is $\frac{1}{100}$ of 500, which is 5. So each 5% is 5 times 5, which is 25. For convenience, mark the "Part" number line in increments of 25 and the "Percent" number line in corresponding increments of 5%.

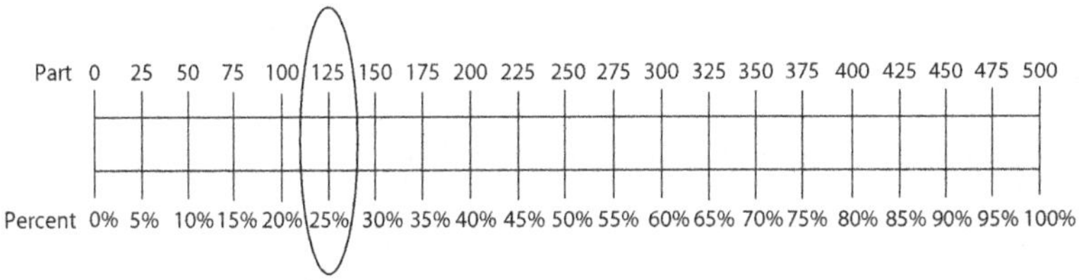

3. The percent is 15% and the whole is 80. The part is

$$15\% \text{ times } 80 = \frac{15}{100} \times 80 = \frac{3}{{}_1\cancel{20}} \times \frac{\cancel{80}^{4}}{1} = \frac{12}{1} = 12$$

There are 12 vans in the parking lot.

4. This question requires two steps. First, find 10% of \$399. Next, subtract the result of Step 1 from \$399.

Step 1. Find 10% of \$399.

$$10\% \text{ of } \$399 = \frac{10}{100} \times \$399 = \frac{1}{10} \times \frac{\$399}{1} = \frac{\$399}{10} = \$39.90$$

Step 2. Subtract \$39.90 from \$399.

$$\$399.00 - \$39.90 = \$359.10$$

The sale price of the recorder is \$359.10.

Finding the Whole, Given the Percent and the Part

If you know the percent and the part, you can find the whole by using equivalent ratios. For instance, if the percent is 30%, $30\% = \frac{30}{100} = \frac{\text{part}}{\text{whole}}$. To find the whole, make the $\frac{\text{part}}{\text{whole}}$ ratio equivalent to $\frac{30}{100}$. Here is an example.

Thirty percent of the students in the stadium are girls. There are 120 girls in the stadium. How many students are in the stadium?

The ratio of girls to students in the stadium is 30%. This fact means the ratio of girls to students is $\frac{30}{100}$. Therefore, $\frac{30}{100}=\frac{\text{part}}{\text{whole}}$. There are 120 girls in the stadium, so the part is 120. The whole is the number that makes the two ratios, $\frac{30}{100}$ and $\frac{120}{\text{whole}}$, equivalent. Think: *"I want to get from 30 to 120. Because 120 is greater than 30, I will need to find a number to multiply 30 by to get 120."* You can find the number by guessing and checking, or you can simply divide 120 by 30 to find the number, like this.

$$\begin{array}{r} 4 \\ 30\overline{)120} \\ \underline{120} \\ 0 \end{array}$$

Now multiply both the numerator and denominator of $\frac{30}{100}$ by 4: $\frac{30\times4}{100\times4}=\frac{120}{400}$. Therefore, $\frac{120}{400}=\frac{120}{\text{whole}}$, which means the whole is 400. There are 400 students in the stadium.

It is usually helpful to use the value of the percent ratio when finding the whole. The **value** of a percent ratio is its equivalent fraction in simplest form. The following double-number-line diagram illustrates some common equivalencies.

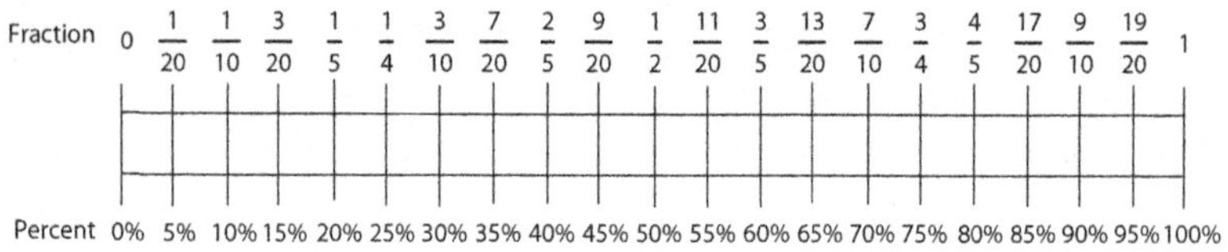

Here is an example.

> Forty percent of the students in the band are boys. There are 18 boys in the band. How many students are in the band?

The ratio of boys to students in the band is 40%. This fact means the ratio of boys to students is $\frac{40}{100}=\frac{2}{5}$. Therefore, $\frac{2}{5}=\frac{\text{part}}{\text{whole}}$. There are 18 boys in the band, so the part is 18. The whole is the number that makes the two ratios, $\frac{2}{5}$ and $\frac{18}{\text{whole}}$, equivalent. Think: *"I want to get from 2 to 18. Because 18 is greater than 2, I will need to find a number to multiply 2 by to get 18. The number I need is 9."*

Now multiply both the numerator and denominator of $\frac{2}{5}$ by 9: $\frac{2\times9}{5\times9}=\frac{18}{45}$. Therefore, $\frac{18}{45}=\frac{18}{\text{whole}}$, which means the whole is 45. There are 45 students in the band.

☞ Try These

1. Flaxseed is 15% of a mixture. The amount of flaxseed is 45 ounces. How many ounces is the mixture?
2. Diaval has driven 75 miles. This distance is 25% of the total distance Diaval will be driving. What is the total distance?
3. Linda saved 20% on the purchase of a desk lamp. She saved $30 off the original price of the desk lamp. What was the original price of the lamp?
4. Seventy percent of the people in the school supply store are teachers. There are 14 teachers in the store. How many people are in the store?

Solutions

1. The percent is $15\% = \frac{15}{100}$. The part is 45 ounces.

   $\frac{15 \times 3}{100 \times 3} = \frac{45}{300} = \frac{\text{part}}{\text{whole}}$. The mixture is 300 ounces.

2. The percent is $25\% = \frac{25}{100} = \frac{1}{4}$. The part is 75 miles.

   $\frac{1 \times 75}{4 \times 75} = \frac{75}{300} = \frac{\text{part}}{\text{whole}}$. The total distance is 300 miles.

3. The percent is $20\% = \frac{20}{100} = \frac{1}{5}$. The part is $30.

   $\frac{1 \times 30}{5 \times 30} = \frac{30}{150} = \frac{\text{part}}{\text{whole}}$. The original price was $150.

4. The percent is $70\% = \frac{70}{100} = \frac{7}{10}$. The part is 14.

   $\frac{7 \times 2}{10 \times 2} = \frac{14}{20} = \frac{\text{part}}{\text{whole}}$. There are 20 people in the store.

2. The Number System

In this chapter, you extend your understanding of arithmetic operations to include division of fractions and operations on multi-digit decimals. The number line is extended to include negative fractions and decimals.

Dividing Fractions by Fractions

(CCSS.Math.Content.6.NS.A.1, CCSS.Math.Content.6.NS.B.4)

Division by fractions includes dividing a whole number by a fraction, dividing fractions that have like denominators, and dividing fractions that have unlike denominators.

Dividing a Whole Number by a Fraction

How many $\frac{1}{4}$-pound hamburger patties can Shalyn make from 3 pounds of hamburger meat?

This question is asking, "How many $\frac{1}{4}$-pounds are in 3 pounds?" In other words, it is asking, "How many fourths are in 3 wholes?" To find the answer, divide 3 by $\frac{1}{4}$. To make the division easier to understand, change 3 wholes to fourths as shown here. ***Tip:*** Remember, $3 = \frac{3}{1}$.

$$3 \div \frac{1}{4} = \frac{3}{1} \div \frac{1}{4} = \frac{3\times4}{1\times4} \div \frac{1}{4} = \frac{12}{4} \div \frac{1}{4} = 12 \text{ fourths} \div 1 \text{ fourth} = \frac{12 \cancel{\text{fourths}}}{1 \cancel{\text{fourth}}} = 12$$

Tip: You can write a division problem as a fraction. That is, $a \div b = \frac{a}{b}$.

Shalyn can make twelve $\frac{1}{4}$-pound hamburger patties from 3 pounds of hamburger meat.

Here is a number-line model of the solution.

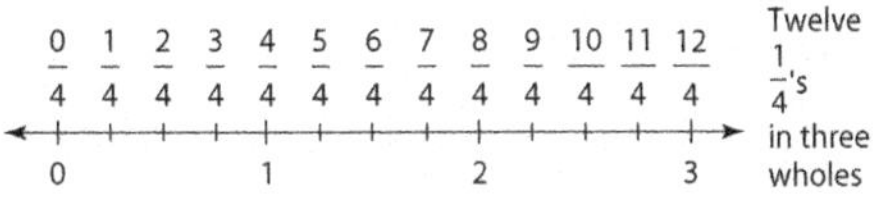

Think *"1 whole is 4 fourths. So 3 wholes are 12 fourths."*

☞ Try These

1. Draw a number-line model to illustrate $3 \div \frac{1}{2} = 6 \text{ halves} \div 1 \text{ half} = \frac{6\ \cancel{\text{halves}}}{1\ \cancel{\text{half}}} = 6$.
2. Compute $5 \div \frac{1}{8}$.
3. Divide 4 by $\frac{1}{3}$.
4. Compute $6 \div \frac{3}{4}$.
5. How many $\frac{3}{8}$-inch strips can be cut from a wire 3 inches long?

Solutions

1.

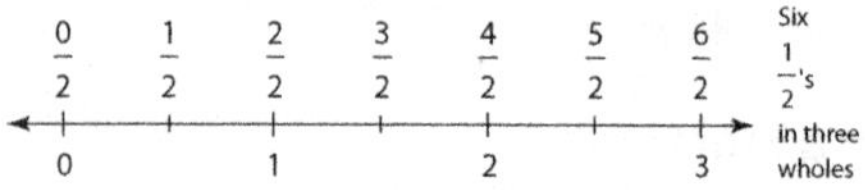

2. $5 \div \frac{1}{8} = \frac{5}{1} \div \frac{1}{8} = \frac{5 \times 8}{1 \times 8} \div \frac{1}{8} = \frac{40}{8} \div \frac{1}{8} = 40 \text{ eighths} \div 1 \text{ eighth} = \frac{40\ \cancel{\text{eighths}}}{1\ \cancel{\text{eighth}}} = 40$
3. $4 \div \frac{1}{3} = \frac{4}{1} \div \frac{1}{3} = \frac{4 \times 3}{1 \times 3} \div \frac{1}{3} = \frac{12}{3} \div \frac{1}{3} = 12 \text{ thirds} \div 1 \text{ third} = \frac{12\ \cancel{\text{thirds}}}{1\ \cancel{\text{third}}} = 12$
4. $6 \div \frac{3}{4} = \frac{6}{1} \div \frac{3}{4} = \frac{6 \times 4}{1 \times 4} \div \frac{3}{4} = \frac{24}{4} \div \frac{3}{4} = 24 \text{ fourths} \div 3 \text{ fourths} = \frac{24\ \cancel{\text{fourths}}}{3\ \cancel{\text{fourths}}} = 8$
5. $3 \div \frac{3}{8} = \frac{3}{1} \div \frac{3}{8} = \frac{3 \times 8}{1 \times 8} \div \frac{3}{8} = \frac{24}{8} \div \frac{3}{8} = 24 \text{ eighths} \div 3 \text{ eighths} = \frac{24\ \cancel{\text{eighths}}}{3\ \cancel{\text{eighths}}} = 8$

   Eight $\frac{3}{8}$-inch strips can be cut from a wire 3 inches long.

Dividing Fractions (Like Denominators)

How many $\frac{1}{3}$-pound hamburger patties can Shalyn make from $2\frac{2}{3}$ pounds of hamburger meat? This question is asking, "How many $\frac{1}{3}$-pounds are in $2\frac{2}{3}$ pounds?" You know $2\frac{2}{3}$ can be represented by $\frac{8}{3}$. So the question is really asking, "How many thirds are in 8 thirds?" To find the answer, divide $\frac{8}{3}$ by $\frac{1}{3}$.

$$\frac{8}{3} \div \frac{1}{3} = 8 \text{ thirds} \div 1 \text{ third} = \frac{8\ \cancel{\text{thirds}}}{1\ \cancel{\text{third}}} = 8$$

Shalyn can make eight $\frac{1}{3}$-pound hamburger patties from $2\frac{2}{3}$ pounds of hamburger meat.

Here is a number-line model of $\frac{8}{3} \div \frac{1}{3} = 8$.

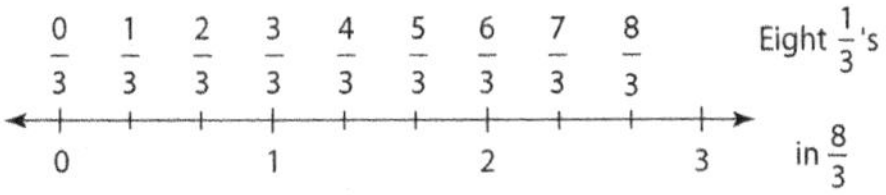

In general, when you divide two fractions that have like denominators, you can use the rule $\frac{a}{d} \div \frac{b}{d} = \frac{a}{b}$. Here are examples.

$$2\frac{1}{4} \div \frac{3}{4} = \frac{9}{4} \div \frac{3}{4} = \frac{9}{3} = 3$$

$$\frac{2}{7} \div \frac{5}{7} = \frac{2}{5}$$

Tip: Be careful when dividing. The order of the numbers in a division problem is important. For instance, $\frac{3}{8} \div \frac{5}{8} = \frac{3}{5}$, but $\frac{5}{8} \div \frac{3}{8} = \frac{5}{3}$.

You will find it useful to keep the following in mind when you divide any two numbers:

- If the dividend is greater than the divisor, the quotient is greater than 1 and tells how many times larger the dividend is than the divisor. For instance, $\frac{10}{3} \div \frac{2}{3} = 5$ means $\frac{10}{3}$ is 5 times larger than $\frac{2}{3}$.
- If the dividend is less than the divisor, the quotient is less than 1 and tells what fraction the dividend is of the divisor. For instance, $\frac{3}{4} \div \frac{9}{4} = \frac{3}{9} = \frac{1}{3}$ means $\frac{3}{4}$ is $\frac{1}{3}$ of $\frac{9}{4}$.

☞ Try These

1. Draw a number-line model to illustrate $\frac{6}{7} \div \frac{2}{7}$ is $\frac{6}{2} = 3$.
2. Fill in the blank.
   - **(a)** $\frac{12}{13} \div \frac{3}{13} =$ ________
   - **(b)** Is your answer to part (a) greater than 1 or less than 1? Explain why.
   - **(c)** $1\frac{1}{9} \div \frac{5}{9} =$ ________
   - **(d)** Is your answer to part (c) greater than 1 or less than 1? Explain why.
   - **(e)** $3\frac{3}{4} \div 1\frac{1}{4} =$ ________
   - **(f)** Is your answer to part (e) greater than 1 or less than 1? Explain why.
   - **(g)** $\frac{2}{7} \div \frac{6}{7} =$ ________
   - **(h)** Is your answer to part (g) greater than 1 or less than 1? Explain why.

3. How many $\frac{3}{4}$-cup servings are in $2\frac{1}{4}$ cups of cooked rice?

4. Jared is filling small bags with $\frac{2}{3}$ ounces of mixed spices. He has $5\frac{1}{3}$ ounces of mixed spices. How many bags can he fill?

5. Fill in the blanks.

   (a) $\frac{8}{9} \div \frac{2}{9} =$ __________. This result means __________ is __________ times as large as __________.

   (b) $\frac{6}{11} \div \frac{12}{11} =$ __________. This result means __________ is __________ of __________.

   (c) $2\frac{1}{2} \div \frac{1}{2} =$ __________. This result means __________ is __________ times as large as __________.

   (d) $\frac{1}{2} \div 2\frac{1}{2} =$ __________. This result means __________ is __________ of __________.

Solutions

1.

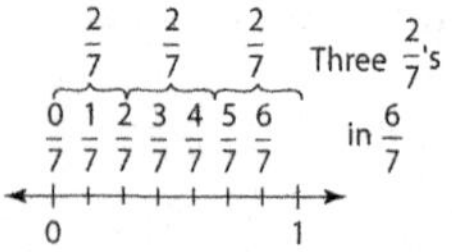

2. **(a)** $\frac{12}{13} \div \frac{3}{13} = \frac{12}{3} = 4$

   (b) The answer is greater than 1 because $\frac{12}{13}$ is greater than $\frac{3}{13}$.

   (c) $1\frac{1}{9} \div \frac{5}{9} = \frac{10}{9} \div \frac{5}{9} = \frac{10}{5} = 2$

   (d) The answer is greater than 1 because $1\frac{1}{9}$ is greater than $\frac{5}{9}$.

   (e) $3\frac{3}{4} \div 1\frac{1}{4} = \frac{15}{4} \div \frac{5}{4} = \frac{15}{5} = 3$

   (f) The answer is greater than 1 because $3\frac{3}{4}$ is greater than $1\frac{1}{4}$.

   (g) $\frac{2}{7} \div \frac{6}{7} = \frac{2}{6} = \frac{1}{3}$

   (h) The answer is less than 1 because $\frac{2}{7}$ is less than $\frac{6}{7}$.

3. You need to find how many $\frac{3}{4}$ cups are in $2\frac{1}{4}$ cups. To find the answer, divide $2\frac{1}{4}$ by $\frac{3}{4}$.

$$2\frac{1}{4} \div \frac{3}{4} = \frac{9}{4} \div \frac{3}{4} = \frac{9}{3} = 3$$

   There are three $\frac{3}{4}$-cup servings in $2\frac{1}{4}$ cups.

4. You need to find how many $\frac{2}{3}$ ounces are in $5\frac{1}{3}$ ounces. To find the answer, divide $5\frac{1}{3}$ by $\frac{2}{3}$.

$$5\frac{1}{3} \div \frac{2}{3} = \frac{16}{3} \div \frac{2}{3} = \frac{16}{2} = 8$$

   Jared can fill 8 bags.

5. **(a)** $\frac{8}{9} \div \frac{2}{9} = \frac{8}{2} = 4$; $\frac{8}{9}$; 4; $\frac{2}{9}$

(b) $\frac{6}{11} \div \frac{12}{11} = \frac{6}{12} = \frac{1}{2}$; $\frac{6}{11}$; $\frac{1}{2}$; $\frac{12}{11}$

(c) $2\frac{1}{2} \div \frac{1}{2} = \frac{5}{2} \div \frac{1}{2} = \frac{5}{1} = 5$; $2\frac{1}{2}$; 5; $\frac{1}{2}$

(d) $\frac{1}{2} \div 2\frac{1}{2} = \frac{1}{2} \div \frac{5}{2} = \frac{1}{5}$; $\frac{1}{2}$; $\frac{1}{5}$; $2\frac{1}{2}$

Dividing Fractions (Unlike Denominators)

Suppose you want to know how many $\frac{1}{6}$-foot pieces can be cut from a board that is $\frac{2}{3}$ foot long. To answer the question, divide $\frac{2}{3}$ by $\frac{1}{6}$. You have two ways to perform the division.

Method 1: First, find a common denominator for the two fractions. A **common denominator** is a number that both denominators will divide evenly into. Next, write the dividend and divisor as equivalent fractions that have the common denominator. Then divide the numerators of the two equivalent fractions. Using this method,

$$\frac{2}{3} \div \frac{1}{6} = \frac{2 \times 2}{3 \times 2} \div \frac{1}{6} = \frac{4}{6} \div \frac{1}{6} = \frac{4}{1} = 4$$

Method 2: Multiply the dividend by the reciprocal of the divisor. The **reciprocal** of a fraction is the fraction made by interchanging the numerator and denominator. In general, $\frac{a}{b} \div \frac{c}{d} = \frac{a}{b} \times \frac{d}{c} = \frac{ad}{bc}$. ***Tip:*** A helpful way to remember this method is to think: *"Keep, change, flip."* You "keep" the first fraction, you "change" division to multiplication, and you "flip" the second fraction to its reciprocal. Using this method,

$$\frac{2}{3} \div \frac{1}{6} = \frac{2}{3} \times \frac{6}{1} = \frac{12}{3} = 4$$

Whether you use Method 1 or Method 2, you get the same the answer: Four $\frac{1}{6}$-foot pieces can be cut from a board that is $\frac{2}{3}$ foot long.

Here is a number-line model of $\frac{2}{3} \div \frac{1}{6} = 4$.

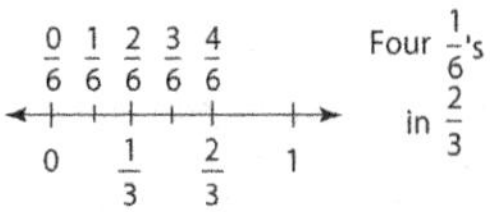

Think: *"1 third is 2 sixths. So 2 thirds is 4 sixths."*

☞ Try These

1. Draw a number-line model to illustrate $\frac{3}{4} \div \frac{3}{8}$ is $\frac{6}{8} \div \frac{3}{8} = \frac{6}{3} = 2$.
2. Find the quotient using Method 1 for dividing fractions with unlike denominators.
   (a) $1\frac{1}{2} \div \frac{3}{4}$
   (b) $\frac{3}{10} \div 1\frac{1}{5}$
3. Find the quotient using Method 2 for dividing fractions with unlike denominators.
   (a) $\frac{3}{4} \div \frac{3}{8}$
   (b) $1\frac{3}{4} \div 3\frac{1}{2}$
4. How many one-half cup servings are in $3\frac{3}{4}$ cups of pudding?

Solutions

1.

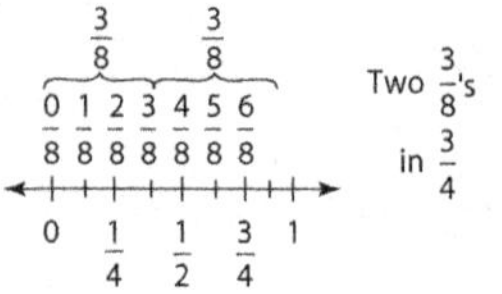

2. **(a)** $1\frac{1}{2} \div \frac{3}{4} = \frac{3}{2} \div \frac{3}{4} = \frac{3\times 2}{2\times 2} \div \frac{3}{4} = \frac{6}{4} \div \frac{3}{4} = \frac{6}{3} = 2$

   (b) $\frac{3}{10} \div 1\frac{1}{5} = \frac{3}{10} \div \frac{6}{5} = \frac{3}{10} \div \frac{6\times 2}{5\times 2} = \frac{3}{10} \div \frac{12}{10} = \frac{3}{12} = \frac{1}{4}$

3. **(a)** $\frac{3}{4} \div \frac{3}{8} = \frac{3}{4} \times \frac{8}{3} = \frac{\cancel{3}^{1}}{_{1}\cancel{4}} \times \frac{\cancel{8}^{2}}{\cancel{3}_{1}} = \frac{2}{1} = 2$

   (b) $1\frac{3}{4} \div 3\frac{1}{2} = \frac{7}{4} \div \frac{7}{2} = \frac{7}{4} \times \frac{2}{7} = \frac{^{1}\cancel{7}}{_{2}\cancel{4}} \times \frac{\cancel{2}^{1}}{\cancel{7}_{1}} = \frac{1}{2}$

4. $3\frac{3}{4} \div \frac{1}{2} = \frac{15}{4} \div \frac{1}{2} = \frac{15}{4} \times \frac{2}{1} = \frac{15}{_{2}\cancel{4}} \times \frac{\cancel{2}^{1}}{1} = \frac{15}{2} = 7\frac{1}{2}$

   There are $7\frac{1}{2}$ one-half cup servings in $3\frac{3}{4}$ cups of pudding.

Performing Operations with Multi-Digit Decimals Using Standard Algorithms

(CCSS.Math.Content.6.NS.B.2, CCSS.Math.Content.6.NS.B.3)

A **standard algorithm** is a well-known, traditional, step-by-step procedure for performing a task. Adding, subtracting, multiplying, and dividing multi-digit decimals using standard algorithms are important skills you should learn to do smoothly and efficiently.

Adding and Subtracting Multi-Digit Decimals

Before you add or subtract decimals, make sure they all have the same number of decimal places. Decimals that have the same number of decimal places are **like decimals.** They have the same fractional unit, so adding or subtracting them makes mathematical sense.

Standard Algorithm for Adding Two or More Decimals

1. Write the numbers one above the other, being sure to line up the decimal points vertically.
2. Fill in empty decimal places with zeroes to ensure all the numbers are like decimals.
3. Add as you would with whole numbers, ignoring the decimal points.
4. Place the decimal point in the answer directly under the decimal points in the problem.

Here is an example.

Compute 103.625 + 145 + 40.05 + 3,875.8.

1. Write the numbers one above the other, being sure to line up the decimal points vertically. ***Tip:*** In a whole number, the decimal point is understood to be to the immediate right of the rightmost digit.

$$\begin{array}{r} 103.625 \\ 145. \\ 40.05 \\ +\ 3{,}875.8 \\ \hline \end{array}$$

2. Fill in empty decimal places with zeroes to ensure all the numbers are like decimals.

$$\begin{array}{r} 103.625 \\ 145.000 \\ 40.050 \\ +3{,}875.800 \\ \hline \end{array}$$

3. Add as you would with whole numbers, ignoring the decimal points.

$$\begin{array}{r} 103.625 \\ 145.000 \\ 40.050 \\ +3{,}875.800 \\ \hline 4{,}164\ 475 \end{array}$$

4. Place the decimal point in the answer directly under the decimal points in the numbers being added.

$$\begin{array}{r} 103.625 \\ 145.000 \\ 40.050 \\ +3{,}875.800 \\ \hline 4{,}164.475 \end{array}$$

So, 103.625 + 145 + 40.05 + 3,875.8 = 4,164.475.

Standard Algorithm for Subtracting Two Decimals

1. Write the minuend above the subtrahend, being sure to line up the decimal points vertically. ***Tip:*** Recall "minuend – subtrahend = difference" in a subtraction problem.
2. Fill in empty decimal places with zeroes to ensure the two numbers are like decimals.
3. Subtract as you would with whole numbers, ignoring the decimal points.
4. Place the decimal point in the answer directly under the decimal points in the problem.

Here is an example.

Compute 200.04 – 32.006.

1. Write the minuend above the subtrahend, being sure to line up the decimal points vertically.

$$\begin{array}{r} 200.04 \\ \underline{-\ 32.006} \end{array}$$

2. Fill in empty decimal places with zeroes to ensure all the numbers are like decimals.

$$\begin{array}{r} 200.040 \\ -\ 32.006 \\ \hline \end{array}$$

3. Subtract as you would with whole numbers, ignoring the decimal points.

$$\begin{array}{r} 200.040 \\ -\ 32.006 \\ \hline 168\ 034 \end{array}$$

4. Place the decimal point in the answer directly under the decimal points in the numbers being added.

$$\begin{array}{r} 200.040 \\ -32.006 \\ \hline 168.034 \end{array}$$

So, 200.04 – 32.006 = 168.034.

☞ Try These

1. Add 65.3, 0.34, and 7.008.
2. Compute 36.814 + 0.56 + 15.
3. Subtract 4.075 from 12.1.
4. Compute 200 – 13.845.
5. A watch is marked \$49. Finn buys the watch on sale and saves \$4.90. He pays \$3.64 in sales tax. How much money did the watch cost plus tax?

Solutions

1. $$\begin{array}{r} 65.300 \\ 0.340 \\ +\ 7.008 \\ \hline 72.648 \end{array}$$

2. $$\begin{array}{r} 36.814 \\ 0.560 \\ +15.000 \\ \hline 52.374 \end{array}$$

3. $$\begin{array}{r} 12.100 \\ -4.075 \\ \hline 8.025 \end{array}$$

4. $$\begin{array}{r} 200.000 \\ -13.845 \\ \hline 186.155 \end{array}$$

5. The sale price of the watch is

$$\begin{array}{r} \$49.00 \\ -\$4.90 \\ \hline \$44.10 \end{array}$$

The cost plus tax is

$$\begin{array}{r} \$44.10 \\ +\$3.64 \\ \hline \$47.74 \end{array}$$

Multiplying Multi-Digit Decimals

As you did when adding and subtracting, multiply decimals just like you multiply whole numbers, except be sure to place the decimal point correctly. When you multiply decimals, there are as many decimal places in the product as there are in all the factors combined.

Standard Algorithm for Multiplying Two Decimals

1. Multiply the two numbers as you would with whole numbers.
2. Count the number of decimal places in each of the numbers being multiplied.
3. Place the decimal point in the proper place in the product. The number of decimal places in the product is the sum of the number of decimal places in the numbers being multiplied. If there are not enough places, insert one or more zeroes as needed to the left of the leftmost nonzero digit.

Here is an example.

Compute 55.7×0.25.

1. Multiply the two numbers as you would with whole numbers.

$$\begin{array}{r} 55.7 \\ \underline{\times\ 0.25} \\ 13.925 \end{array}$$

2. Count the number of decimal places in each of the numbers being multiplied. ***Tip:*** Start at the right and work left in counting decimal places.

$$\begin{array}{rl} 55.7 & \text{(1 place)} \\ \underline{\times\ 0.25} & \text{(2 places)} \\ 13.925 & \end{array}$$

3. Place the decimal point in the proper place in the product. The number of decimal places in the product is the sum of the number of decimal places in the numbers being multiplied. If there are not enough places, insert one or more zeroes as needed to the left of the leftmost nonzero digit.

$$\begin{array}{rr} 55.7 & \text{(1 place)} \\ \underline{\times\ 0.25} & \underline{\text{(+ 2 places)}} \\ 13.925 & \text{(3 places)} \end{array}$$

So, $55.7 \times 0.25 = 13.925$.

☞ Try These

1. Multiply 34.5 by 0.643.
2. Compute 40.564×65.
3. Find the product of 0.006 and 4.319.
4. What is 0.002 times 0.0003?
5. A regular-size drum set costs $195.50. A junior-size drum set is 0.62 as much. What is the cost of the junior-size drum set?

Solutions

1. $\begin{array}{r} 34.5 \\ \times\ 0.643 \\ \hline 22.1835 \end{array}$

2. $\begin{array}{r} 40.564 \\ \times\ 65 \\ \hline 2636.660 \end{array}$

3. $\begin{array}{r} 0.006 \\ \times\ 4.319 \\ \hline 0.025914 \end{array}$

4. $\begin{array}{r} 0.002 \\ \times\ 0.0003 \\ \hline 0.0000006 \end{array}$

5. $\begin{array}{r} 195.50 \\ \times\ 0.62 \\ \hline 121.2100 \end{array}$

   The cost of the junior-size drum set is $121.21.

Dividing Multi-Digit Decimals

When you use long division to divide a decimal by a whole number, place the decimal point in the quotient directly above the decimal point in the dividend. When the divisor is a decimal, first multiply both dividend and divisor by the power of 10 (10, 100, 1,000, and so on) to make the divisor a whole number. Then divide.

Algorithm for Dividing Two Decimals Using the Equivalent Fraction Form

1. Rewrite dividend ÷ divisor or $\text{divisor}\overline{\smash{)}\text{dividend}}$ in equivalent fraction form as $\frac{\text{dividend}}{\text{divisor}}$.
2. Multiply both the dividend and divisor of the equivalent fraction by the power of 10 (10, 100, 1,000, and so forth) to make the divisor a whole number. ***Tip***: Multiplying $\frac{\text{dividend}}{\text{divisor}}$ by $\frac{10}{10}, \frac{100}{100}, \frac{1{,}000}{1{,}000}$, and so forth is equivalent to multiplying by 1. So the value of $\frac{\text{dividend}}{\text{divisor}}$ remains the same.
3. Using long division, divide as you would with whole numbers, ignoring the decimal point.
4. Place the decimal point in the quotient directly above the decimal point in the dividend.

Here is an example.

Compute $0.002\overline{)2.4826}$.

1. Rewrite $\text{divisor}\overline{)\text{dividend}}$ in equivalent fraction form as $\frac{\text{dividend}}{\text{divisor}}$.

$$0.002\overline{)2.4826} = \frac{2.4826}{0.002}$$

2. Multiply both the dividend and divisor of the equivalent fraction by the power of 10 (10, 100, 1,000, and so forth) to make the divisor a whole number.

$$\frac{2.4826}{0.002} = \frac{2.4826 \times 1{,}000}{0.002 \times 1{,}000} = \frac{2482.6}{2}$$

3. Using long division, divide as you would with whole numbers, ignoring the decimal point.

$$\begin{array}{r} 1241\ 3 \\ 2\overline{)2482.6} \end{array}$$

4. Place the decimal point in the quotient directly above the decimal point in the dividend.

$$\begin{array}{r} 1241.3 \\ 2\overline{)2482.6} \end{array}$$

So, the answer to $0.002\overline{)2.4826}$ is 1,241.3.

Once you comfortably understand the mathematical reasoning in the above algorithm—that lets you work the problem $2\overline{)2482.6}$ to get the answer to the problem $0.002\overline{)2.4826}$—you can shorten the process using the following algorithm. This second algorithm is most commonly referred to as the "standard algorithm" for dividing decimals.

Standard Algorithm for Dividing Two Decimals

1. Multiply both the dividend and divisor by the power of 10 (10, 100, 1,000, and so forth) to make the divisor a whole number. Do this multiplication mentally. ***Tip:*** To multiply any number by 10, 100, 1,000, and so forth, move its decimal point to the right as many places as there are zeroes in the multiplier. If necessary, attach additional zeroes after the dividend's rightmost digit.
2. Using long division, divide as you would with whole numbers, ignoring the decimal point.
3. Place the decimal point in the quotient directly above the decimal point in the dividend.

Here is an example.

Compute $1.06\overline{)90.524}$.

1. Multiply both the dividend and divisor by the power of 10 (10, 100, 1,000, and so forth) to make the divisor a whole number. Do this multiplication mentally.

$$1.06\overline{)90.524} = 1.\underrightarrow{06}\overline{)90.\underrightarrow{52}4} = 106\overline{)9052.4}$$

2. Using long division, divide as you would with whole numbers, ignoring the decimal point.

$$\begin{array}{r} 85.4 \\ 106\overline{)\,9052.4} \\ \underline{-848} \\ 572 \\ \underline{-530} \\ 424 \\ \underline{-424} \\ 0 \end{array}$$

3. Place the decimal point in the quotient directly above the decimal point in the dividend.

$$\begin{array}{r} 85.4 \\ 106\overline{)\,9052.4} \\ \underline{-848} \\ 572 \\ \underline{-530} \\ 424 \\ \underline{-424} \\ 0 \end{array}$$

So, the answer to $1.06\overline{)90.524}$ is 85.4.

Tip: If you have a remainder when you divide, you can keep attaching zeroes to the dividend after its rightmost digit to the right of the decimal point to carry the division to as many decimal places as you wish.

Try These

1. Find the quotient using the equivalent fraction form of the problem.
   - **(a)** $0.00228 \div 1.14$
   - **(b)** $36 \div 0.03$

2. Find the quotient using the standard algorithm.
   - **(a)** $102.75 \div 82.2$
   - **(b)** Divide 3.1512 by 0.026.
   - **(c)** Compute $1.001\overline{)3003}$.
   - **(d)** Find the quotient of 9.2 and 6.25.
   - **(e)** Compute $\dfrac{19.208}{1.4}$.

3. Maipura needs 10 pieces of yarn, each 0.5 meter in length. She will cut the pieces from a piece of yarn that is 4.75 meters long. Does she have enough yarn to cut 10 pieces?

Solutions

1. **(a)**

$$0.00228 \div 1.14 = \frac{0.00228}{1.14} = \frac{0.00228 \times 100}{1.14 \times 100} = \frac{0.228}{114} = 114\overline{)0.228}^{\,0.002}$$

$$\begin{array}{r} 0.002 \\ 114\overline{)0.228} \\ -228 \\ \hline 0 \end{array}$$

(b)

$$36 \div 0.03 = \frac{36}{0.03} = \frac{36 \times 100}{0.03 \times 100} = \frac{3,600}{3} = 1,200$$

2. **(a)**

$$82.2\overline{)102.75} = 82.2\overline{)102.75} = 822\overline{)\ 1027.50}$$

$$\begin{array}{r} 1.25 \\ 822\overline{)\ 1027.50} \\ -822 \\ \hline 2055 \\ -1644 \\ \hline 4110 \\ -4110 \\ \hline 0 \end{array}$$

(b)

$$0.026\overline{)3.1512} = 0.026\overline{)3.1512} = 26\overline{)\ 3151.2}$$

$$\begin{array}{r} 121.2 \\ 26\overline{)\ 3151.2} \\ -26 \\ \hline 55 \\ -52 \\ \hline 31 \\ -26 \\ \hline 52 \\ -52 \\ \hline 0 \end{array}$$

(c)

$$1.001\overline{)3003} = 1.001\overline{)3003.000} = 1001\overline{)\ 3003000.}$$

$$\begin{array}{r} 3000. \\ 1001\overline{)\ 3003000.} \\ -3003 \\ \hline 0 \end{array}$$

Tip: For each of the three 0s after the second 3 in the dividend, you will have a 0 in the quotient.

(d)

$$6.25\overline{)9.2} = 6.25\overline{)9.20} = 625\overline{)\ 920.000}$$

$$\begin{array}{r} 1.472 \\ 625\overline{)\ 920.000} \\ -625 \\ \hline 2950 \\ -2500 \\ \hline 4500 \\ -4375 \\ \hline 1250 \\ -1250 \\ \hline 0 \end{array}$$

Tip: You can keep attaching zeroes to the dividend after the decimal point to carry the division to as many decimal places as you wish. For this problem, the remainder is zero after three decimal places.

(e)

$$\frac{19.208}{1.4} = 1.4\overline{)19.208} = 1.\underset{\rightarrow}{4}\overline{)19.\underset{\rightarrow}{2}08} = 14\overline{)\ 192.08}$$

$$\begin{array}{r} 13.72 \\ 14\overline{)\,192.08} \\ \underline{-14} \\ 52 \\ \underline{-42} \\ 100 \\ \underline{-98} \\ 28 \\ \underline{-28} \\ 0 \end{array}$$

3. This question is asking "How many 0.5 meter are in 4.75 meters?" Reason through the question. You know 0.5 is $\frac{1}{2}$. So, the question is really asking "How many halves are in 4.75?" You know there are 8 halves in 4 wholes and 10 halves in 5 wholes. So, Maipura needs 10 meters of yarn. Thus, 4.75 meters is not enough yarn. You can verify your thinking by dividing 4.75 by 0.5.

$$0.5\overline{)4.75} = 0.\underset{\rightarrow}{5}\overline{)4.\underset{\rightarrow}{7}5} = 5\overline{)\ 47.5}$$

$$\begin{array}{r} 9.5 \\ 5\overline{)\,47.5} \\ \underline{-45} \\ 25 \\ \underline{-25} \\ 0 \end{array}$$

The division shows Maipura will be able to cut 9.5 pieces of length 0.5 meter. She does not have enough yarn to cut 10 pieces of length 0.5 meter.

Finding the Greatest Common Factor and Least Common Multiple of Two Whole Numbers

(CCSS.Math.Content.6.NS.C.5)

Two concepts you will find useful to know are greatest common factor and least common multiple.

Finding the Greatest Common Factor of Two Numbers

A **factor** of a number divides evenly into the number with no remainder. The **greatest common factor** (or **GCF**) of two numbers is the largest factor common to the two numbers. Finding the GCF of two numbers is a two-step process.

Step 1. List all the factors of the two numbers.

Step 2. Select the greatest factor common to both.

Here is an example.

Find the GCF of 8 and 12.

Step 1. List all the factors of 8 and all the factors of 12.

The factors of 8 are 1, 2, 4, 8.

The factors of 12 are 1, 2, 3, 4, 6, 12.

Step 2. Examine the two lists and select the greatest factor common to both.

Factors of 8 are 1, 2, $\boxed{4}$, 8

Factors of 12 are 1, 2, 3, $\boxed{4}$, 6, 12

4 is the greatest factor common to both lists.

So, the GCF of 8 and 12 is 4.

Suppose you want to write 8 + 12 as a multiple of a sum of two whole numbers with no common factor. You know the GCF of 8 and 12 is 4. You can factor out the GCF from the two numbers and rewrite the sum using the distributive property, as shown here.

$$8 + 12 = 4(2) + 4(3) = 4(2 + 3)$$

Tip: The distributive property allows you to write $k(a) + k(b)$ as $k(a + b)$.

The GCF also is helpful when you are simplifying fractions. To put a fraction in simplest form, divide its numerator and denominator by the GCF of the numerator and denominator.

Here is an example.

Simplify $\frac{48}{60}$.

To simplify $\frac{48}{60}$, divide its numerator and denominator by the GCF of 48 and 60. Factors of 48 are 1, 2, 3, 4, 6, 8, $\boxed{12}$, 16, 24, 48. Factors of 60 are 1, 2, 3, 4, 5, 6, 10, $\boxed{12}$, 15, 20, 30, 60. So, the GCF is 12. Then, $\frac{48}{60} = \frac{48 \div 12}{60 \div 12} = \frac{4}{5}$.

Tip: The GCF of two numbers is also known as their *greatest common divisor*. This terminology is logical because the GCF is the greatest number that will divide evenly into both numbers.

☞ Try These

1. Find the GCF of 20 and 15.
2. What is the GCF of 18 and 23?

3. Write the fraction $\frac{63}{84}$ in simplest form.
4. Express 30 + 36 as a multiple of a sum of two whole numbers with no common factors.
5. A farmer wants to partition a rectangular 24-foot by 36-foot field into smaller square plots of land. What is the length of the side of the largest square plots into which the farmer can partition the field so that no land is left over?

Solutions

1. Factors of 20 are 1, 2, 4, $\boxed{5}$, 10, 20. Factors of 15 are 1, 3, $\boxed{5}$, 15. So, the GCF is 5.
2. 23 is a prime number. Its factors are 1 and 23. So, the GCF of 18 and 23 is 1.
3. To simplify $\frac{63}{84}$, divide its numerator and denominator by the GCF of 63 and 84. Factors of 63 are 1, 3, 7, 9, $\boxed{21}$, 63. Factors of 84 are 1, 2, 3, 4, 6, 7, 12, 14, $\boxed{21}$, 28, 42, 84. So, the GCF is 21. Then,

   $\frac{63}{84} = \frac{63 \div 21}{84 \div 21} = \frac{3}{4}$.
4. First, find the GCF of 30 and 36. Next, factor out the GCF from the two numbers and rewrite the sum using the distributive property.

   Factors of 30 are 1, 2, 3, 5, $\boxed{6}$, 10, 15, 30. Factors of 36 are 1, 2, 3, 4, $\boxed{6}$, 9, 12, 18, 36. So, the GCF is 6. Then, 30 + 36 = 6(5) + 6(6) = 6(5 + 6).
5. Make a sketch.

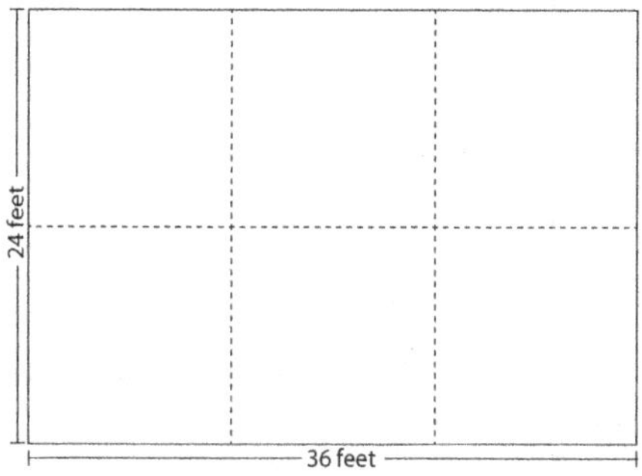

The length of the side of the square plots into which the farmer can partition the field must be a factor of both 24 and 36. The largest such factor is the GCF of 24 and 36.

Factors of 24 are 1, 2, 3, 4, 6, 8, $\boxed{12}$, 24. Factors of 36 are 1, 2, 3, 4, 6, 9, $\boxed{12}$, 18, 36. So, the GCF is 12.

12 feet is the length of the side of the largest square plots into which the farmer can partition the field so that no land is left over.

Finding the Least Common Multiple of Two Numbers

A **multiple** of a number is the product of the number and any whole number. The **least common multiple** (or **LCM**) of two numbers is the least whole number (greater than 0) that is a multiple of both numbers. You can find the LCM of two numbers by listing multiples of the two numbers until you identify the LCM.

Here is an example.

Find the LCM of 6 and 8.

Multiples of 6 are 6, 12, 18, $\boxed{24}$, 30, 36, 42, 48.... Multiples of 8 are 8, 16, $\boxed{24}$, 32, 40, 48.... So, the LCM is 24. Notice $6 \times 8 = 48$ is also a common multiple, but it is not the *least* common multiple.

To simplify this process, start by listing multiples of the greater number. List the greater number, then multiply it by 2, by 3, by 4, and so on. Stop when you first list a multiple that is also a multiple of the other number. This multiple will be the LCM of the two numbers.

Here is an example.

Find the LCM of 9 and 15.

Start by listing multiples of 15→ 15, 30, 45. Stop because $45 = 5 \times 9$, a multiple of 9. So, the LCM of 9 and 15 is 45.

The LCM is helpful when you are adding or subtracting fractions. The LCM of the denominators of the fractions is their least common denominator.

Here is an example.

Compute $\frac{7}{9} + \frac{2}{15}$.

The LCM of 9 and 15 is 45. Change the problem using equivalent fractions with a denominator of 45. Then add as you do when fractions have the same denominator.

$$\frac{7}{9} + \frac{2}{15} = \frac{7 \times 5}{9 \times 5} + \frac{2 \times 3}{15 \times 3} = \frac{35}{45} + \frac{6}{45} = \frac{41}{45}$$

Tip: You can extend the process of finding the LCM to more than two numbers. Start by listing multiples of the greatest number until the first common multiple is found.

☞ Try These

1. Find the LCM of 20 and 15.
2. What is the LCM of 6, 10, and 15?
3. Compute $\frac{3}{10} + \frac{5}{12}$.

4. Elissa gets her hair cut every 15 days, and Jake gets his hair cut every 12 days. How many days will it be before they will get haircuts on the same day?
5. A teacher buys pencils and pens for her classroom. The pencils come 10 in a pack and the pens come 8 in a pack. The teacher buys the same number of pens as she does pencils, with none left over. What is the least number of each she buys?

Solutions

1. Start by listing multiples of 20→ 20, 40, 60. Stop because $60 = 4 \times 15$, a multiple of 15. So, the LCM of 20 and 15 is 60.
2. Start by listing multiples of 15→ 15, 30. Stop because $30 = 3 \times 10$, a multiple of 10, and $30 = 5 \times 6$, a multiple of 6. So, the LCM of 6, 10, and 15 is 30.
3. $\frac{3}{10}+\frac{5}{12}=\frac{3\times 6}{10\times 6}+\frac{5\times 5}{12\times 5}=\frac{18}{60}+\frac{25}{60}=\frac{43}{60}$
4. Elissa's haircut days occur in 15 days, 30 days, 45 days, $\boxed{\text{60 days}}$, 75 days.

   Jake's haircut days occur in 12 days, 24 days, 36 days, 48 days, $\boxed{\text{60 days}}$, 72 days. So, in 60 days Elissa and Jake will get haircuts on the same day. Notice the solution is the LCM of 15 and 12 days.
5. Solve this problem by finding the LCM of 10 and 8. Start by listing multiples of 10→ 10, 20, 30, 40. Stop because $40 = 5 \times 8$, a multiple of 8. So, the LCM of 10 and 8 is 40. The teacher bought 40 pencils (4 packs of 10 in a pack) and 40 pens (5 packs of 8 in a pack).

Understanding the Integers

(CCSS.Math.Content.6.NS.C.6.A, CCSS.Math.Content.6.NS.C.6.C)

The whole numbers are the numbers 0, 1, 2, 3, and so on. You represent the whole numbers as equally spaced points on a number line, increasing endlessly as you move to the right, as shown here.

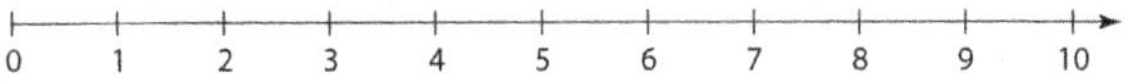

Tip: The intervals on a number line should be equal from one mark to the next.

The numbers to the right of zero on the number line are **positive** numbers. Every positive number has an **opposite.** The opposite of a positive number is a **negative** number. Show the negative numbers by extending the number line to the left. Attach a short horizontal line, called a **negative sign,** to the left of the number to indicate it is a negative number, as shown here.

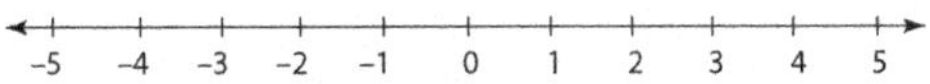

The point one unit to the left of zero on the number line is –1. Read "–1" as "negative one." The point two units to the left of zero is –2. Read "–2" as "negative two." The point three units to the left of zero is –3, and so on. Zero is neither positive nor negative.

Tip: Attaching a + sign to positive numbers is not necessary (although it's not wrong to do so). If no sign is attached, you know the number is positive.

The whole numbers and their opposites are the **integers.** The numbers 1 and –1 are opposites. The numbers 2 and –2 are opposites. The numbers 3 and –3 are opposites, and so on. The number 0 is its own opposite. On the number line, a number and its opposite are the same distance from zero.

Tip: The opposite of a nonzero number can be positive or negative. If a number is positive, its opposite is negative. If a number is negative, its opposite is positive.

You graph a set of numbers by marking a large dot at each point corresponding to one of the numbers. Here is a graph of –4, –1, 0, 3, and 5.

Try These

1. Name the opposite of the given integer.
   (a) 7
   (b) 0
   (c) –6
   (d) 25
   (e) –30

2. Name the set of numbers graphed.
   (a)
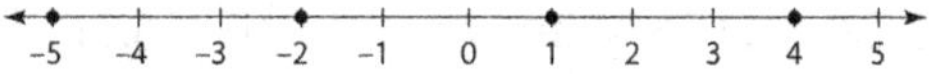

   (b)
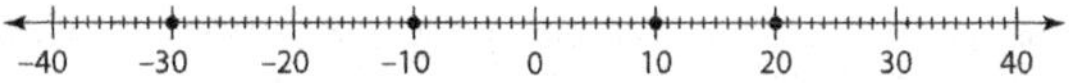

   (c)
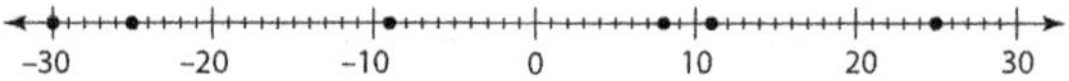

3. Graph each set of numbers on the given number line.
   (a) –3, –2, –1, 0
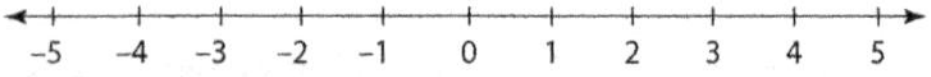

   (b) –15, –5, 0, 5, 15
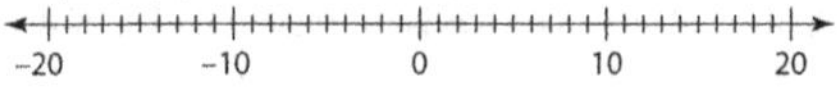

   (c) The opposites of –3, –2, –1, 0
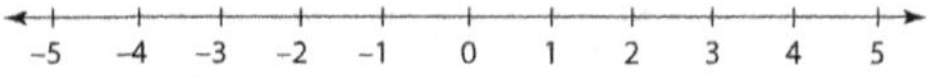

4. Fill in the blank(s).
   - **(a)** The number 8 is 8 units to the __________ of zero on the number line. The opposite of 8 is 8 units to the __________ of zero.
   - **(b)** The number −15 is 15 units to the __________ of zero on the number line. The opposite of −15 is 15 units to the __________ of zero.
   - **(c)** The opposite of the opposite of 5 is __________.
   - **(d)** The opposite of the opposite of −3 is __________.
   - **(e)** The opposite of the opposite of any number *n* is __________.

Solutions

1. **(a)** −7
   (b) 0
   (c) 6
   (d) −25
   (e) 30

2. **(a)** −5, −2, 1, 4
   (b) −30, −10, 11, 20
   (c) −30, −25, −9, 8, 11, 25

3. **(a)**

   (b)

   (c)

4. **(a)** right; left
   (b) left; right
   (c) 5
   (d) −3
   (e) *n*

Understanding the Rational Numbers
(CCSS.Math.Content.6.NS.C.6.A, CCSS.Math.Content.6.NS.C.6.C)

The **rational numbers** are all the numbers that can be written as ratios of two integers, where zero is *not* the denominator of the ratio. In simple terms, the rational numbers include the whole numbers, integers, and all numbers that can be written as positive or negative fractions. The positive rational numbers are to the right of zero on the number line, and the negative rational numbers are to the left of zero on the number line.

Here are examples.

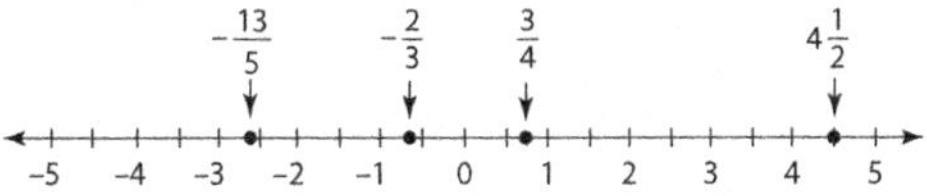

The rational numbers include positive and negative decimals.

Here are examples.

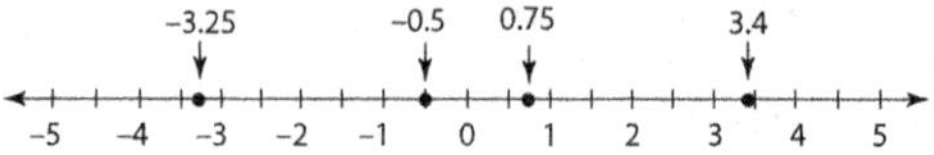

As you might suppose, every rational number has an opposite. If a rational number is positive, its opposite is negative. If a rational number is negative, its opposite is positive. For instance, the numbers $\frac{3}{4}$ and $-\frac{3}{4}$ are opposites. The numbers –3.25 and 3.25 are opposites. The numbers 3.4 and –3.4 are opposites. The numbers –0.5 and $\frac{1}{2}$ are opposites (because $-0.5 = -\frac{5}{10} = -\frac{1}{2}$). The number 0 is its own opposite. As stated before, a number and its opposite are the same distance from zero on the number line.

☞Try These

1. Answer the question.

   (a) Which of the numbers in the following set are rational numbers? $\frac{2}{3}$, -5.23, 0, -1, 100, $-\frac{4}{5}$, 12.9, $\frac{25}{8}$

   (b) Is $\frac{0}{4}$ a rational number? Explain your answer.

   (c) Is $\frac{12}{0}$ a rational number? Explain your answer.

   (d) Which of the following sets of numbers are rational numbers? whole numbers, integers, positive and negative fractions and decimals

2. Name the opposite of the given number.

   (a) 5.3

   (b) –8.69

   (c) $\frac{5}{6}$

   (d) –660

   (e) $-\frac{15}{7}$

   (f) 3,000

3. Name the set of numbers graphed.

   (a) Use decimals: 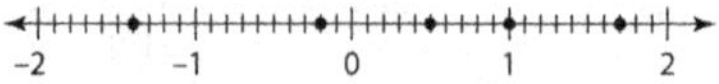

   (b) Use fractions: 

   (c) 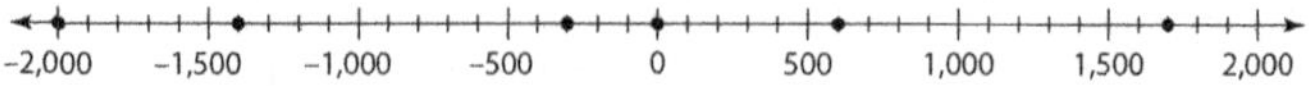

4. Graph each set of numbers on the given number line.

(a) −1.8, −1.2, 0.5, 1.6

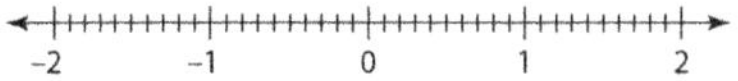

(b) $-2\frac{1}{4}, -1\frac{1}{2}, \frac{3}{4}, \frac{7}{4}$

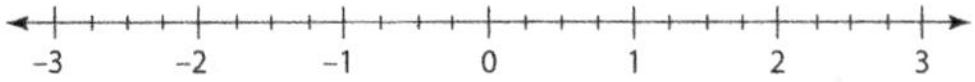

(c) −1,500, −500, 500, 1,500

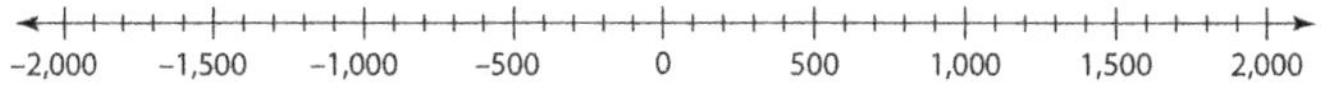

5. Fill in the blank(s).

(a) A rational number is the ratio of two ____________, where zero is *not* the ____________ of the ratio.

(b) The number 9.75 is 9.75 units to the _________ of zero on the number line. The opposite of 9.75 is 9.75 units to the _________ of zero.

(c) The number $-5\frac{3}{8}$ is $5\frac{3}{8}$ units to the _________ of zero on the number line. The opposite of $5\frac{3}{8}$ is $5\frac{3}{8}$ units to the _________ of zero.

(d) The opposite of the opposite of 6.99 is _________.

(e) The opposite of the opposite of $-\frac{24}{5}$ is _________.

Solutions

1. **(a)** All of them

(b) Yes, because $\frac{0}{4} = 0 \div 4 = 0$, which is a rational number.

(c) No, because $\frac{12}{0}$ means $12 \div 0$, which has no meaning. ***Tip:*** Dividing by zero is always wrong.

(d) All of them

2. **(a)** −5.3

(b) 8.69

(c) $-\frac{5}{6}$

(d) 660

(e) $\frac{15}{7}$

(f) −3,000

3. **(a)** –1.4, –0.2, 0.5, 1, 1.7

   (b) $-2\frac{3}{4}, -1\frac{1}{2}, -\frac{3}{4}, \frac{1}{4}, 2\frac{1}{2}$

   (c) –2,000, –1,400, –300, 0, 600, 1,700

4. **(a)** 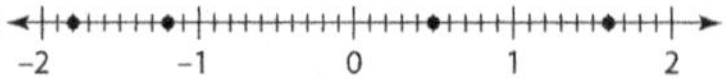

   (b) 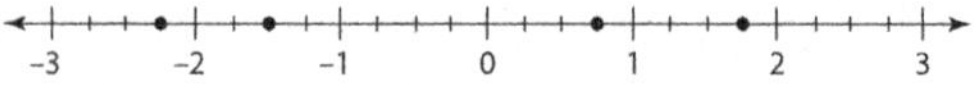

   (c) 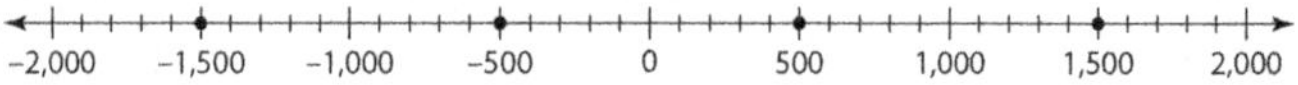

5. **(a)** integers; denominator

   (b) right; left

   (c) left; right

   (d) 6.99

   (e) $-\frac{24}{5}$

Comparing and Ordering Rational Numbers

(CCSS.Math.Content.6.NS.C.6.C, CCSS.Math.Content.6.NS.C.7.A)

The following table summarizes inequality symbols you should know. You use these symbols when comparing and ordering rational numbers.

Common Inequality Symbols

Inequality Symbol	Example	Read the Expression as
$<$	$3 < 8$	3 "is less than" 8
$>$	$4 > 1$	4 "is greater than" 1
$\leq$	$6 \leq 6$	6 "is less than or equal to" 6
$\geq$	$10 \geq 2$	10 "is greater than or equal to" 2
$\neq$	$0 \neq -5$	0 "is not equal to" –5

When you compare two rational numbers, think of their relative locations on the number line. If the numbers have the same location, they are equal. If they don't, they are not equal. Then, the number that is farther to the *right* is the greater number.

Therefore, you immediately know these two facts:

- Negative numbers are always less than zero, and they are less than all the positive numbers.
- Zero is less than all the positive numbers.

Here is an example.

Which is greater: –100 or 20? Support your answer by graphing –100 and 20 on a number line.

20 is greater than −100 because positive numbers are greater than negative numbers and 20 is to the right of −100 on the number line. Here is a graph of −100 and 20.

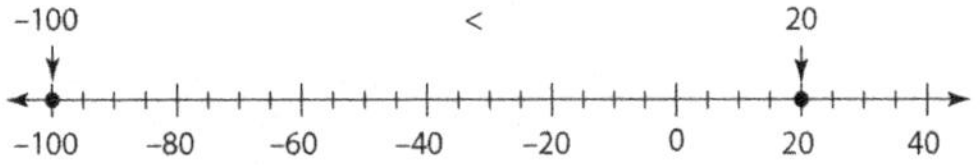

As you move to the right on the number line, the numbers increase in value. When you compare two positive numbers, the one that is "more positive" is the *greater* number. Here is an example.

4 > 1 because as shown on the number line below, 4 lies to the right of 1.

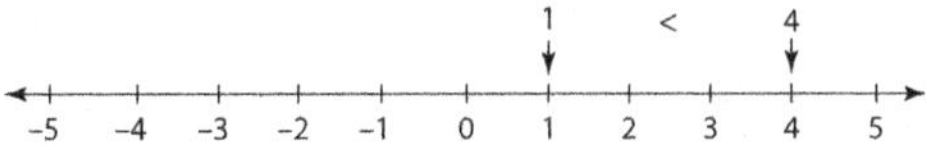

As you move to the left on the number line, the numbers decrease in value. When you compare two negative numbers, the one that is "more negative" is the *lesser* number. Here are examples.

−4 < −1 because as shown on the number line below, −1 lies to the right of −4. Think: *"If I start at zero and move to the left, I will get to −1 before I get to −4, so −1 is the greater number."*

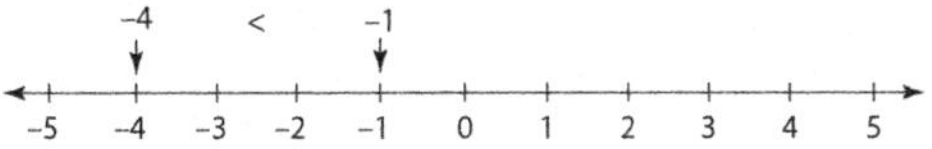

$-\frac{13}{5} < -\frac{2}{3}$ because as shown on the number line below, $-\frac{2}{3}$ lies to the right of $-\frac{13}{5}$. Think: *"If I start at zero and move to the left, I will get to $-\frac{2}{3}$ before I get to $-\frac{13}{5}$, so $-\frac{2}{3}$ is the greater number."*

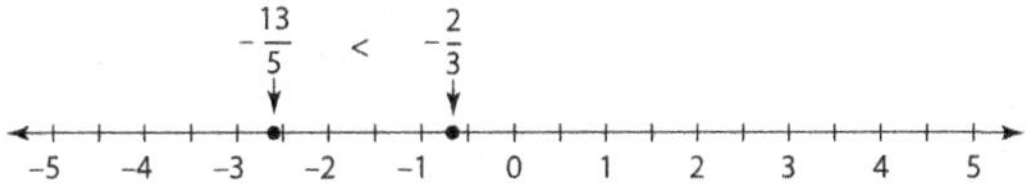

−3.25 < −0.5 because as shown on the number line below, −0.5 lies to the right of −3.25. Think: *"If I start at zero and move to the left, I will get to −0.5 before I get to −3.25, so −0.5 is the greater number."*

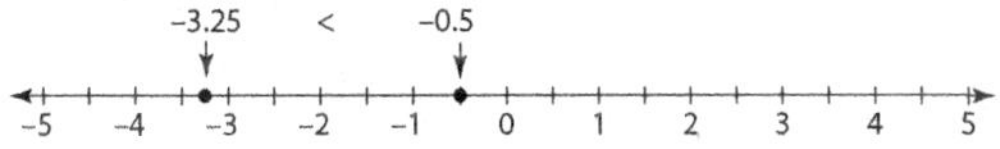

When students first work with positive and negative numbers, some have trouble making sense of comparisons like −3.25 is less than −0.5. What you must always keep in mind is "less than" or "greater than" depends *only* on where the numbers are relative to each other on the number line. When you compare numbers, the number that is least is farthest to the left on the number line, and the number that is greatest is farthest to the right. Practice mentally visualizing numbers on the number line so that comparing rational numbers becomes an easy and automatic skill for you.

When you compare decimal numbers, compare the digits in each place value from left to right. If the decimal numbers do not have the same number of decimal places, attach or delete zeroes after the last digit to the right of the decimal point to make the number of decimal places the same. ***Tip:*** Attaching or deleting zeroes after the last digit to the right of the decimal point does not change the value of a decimal number. For example, 2.8 = 2.80 = 2.800 = 2.8000, and so on.

2.14 < 2.8 because 2.14 < 2.80. The number 2.8 is to the right of 2.14 on the number line. But –2.14 > –2.8 because –2.14 > –2.80. The number –2.14 is to the right of –2.8 on the number line. (See the number line below.)

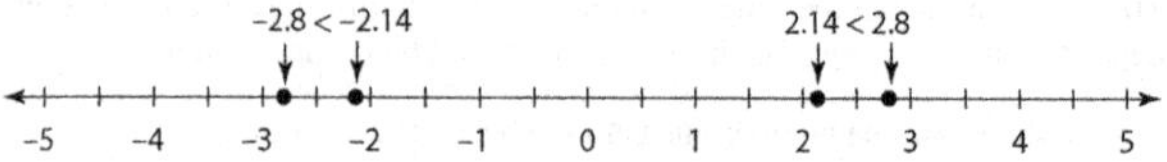

When comparing fractions that have the same denominator, compare the numerators.

For example, $\frac{7}{8} > \frac{3}{8}$ because 7 > 3. And $-\frac{7}{8} < -\frac{3}{8}$ because –7 < –3. (See the number line below.)

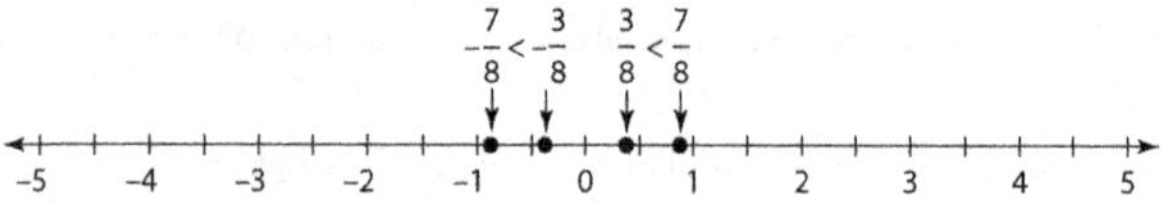

Tip: The above examples illustrate if *a* < *b* it must be true –*a* > –*b*, and if *a* > *b*, it must be true –*a* < –*b*.

When comparing fractions, if the denominators are not the same, write the fractions as equivalent fractions using a common denominator.

For example, to compare $\frac{7}{12}$ and $\frac{5}{8}$, write the fractions as equivalent fractions using the least common denominator.

$$\frac{7}{12} = \frac{7 \times 2}{12 \times 2} = \frac{14}{24} \text{ and } \frac{5}{8} = \frac{5 \times 3}{8 \times 3} = \frac{15}{24}$$

So $\frac{7}{12} < \frac{5}{8}$ because $\frac{14}{24} < \frac{15}{24}$.

To compare a decimal and a fraction, change the fraction to a decimal by dividing its numerator by its denominator. Then compare the two decimals.

For instance, to compare 0.78 and $\frac{7}{8}$, first rewrite $\frac{7}{8}$ as an equivalent decimal.

$$\frac{7}{8} = 7 \div 8 = 8\overline{)\,7.000} = 0.875$$

$$\begin{array}{r} 0.875 \\ 8\overline{)\,7.000} \\ \underline{-64} \\ 60 \\ \underline{-56} \\ 40 \\ \underline{-40} \\ 0 \end{array}$$

So $0.78 < \frac{7}{8}$ because 0.780 < 0.875.

To put a mixture of decimals and fractions in order, change the fractions to decimals. Round them off if they repeat. Then compare the decimal numbers. Before you do, check whether they have the same number of decimal places. If they do not, then change them to make the number of decimal places the same. Here is an example.

Order the numbers $\frac{7}{8}, \frac{2}{3}, -0.35, -\frac{1}{2}, -4.8, 0, -4.58$, and 0.9 from least to greatest.

Write $\frac{7}{8}$ as a decimal by dividing: $7 \div 8 = 0.875$. Similarly, write $\frac{2}{3}$ as 0.667 (rounded to 3 places). Write −0.35 as −0.350. Write $-\frac{1}{2}$ as −0.500. Write −4.8 as −4.800, −4.58 as −4.580, and 0.9 as 0.900. So now you have the following equivalent list of the original numbers.

$$0.875, 0.667, -0.350, -0.500, -4.800, 0, -4.580, 0.900$$

Order these numbers.

$$-4.800, -4.580, -0.500, -0.350, 0, 0.667, 0.875, 0.900$$

Now replace the "stand-ins" with the original numbers to obtain your final answer.

$$-4.8,\ -4.58,\ -\frac{1}{2},\ -0.35, 0, \frac{2}{3}, \frac{7}{8}, 0.9$$

Try These

1. Replace the $\square$ with <, >, or = to make a true statement.

   (a) $-6.7 \square -6.74$

   (b) $8.63 \square 8.595$

   (c) $-\frac{9}{11} \square -\frac{7}{11}$

   (d) $-0.24 \square -0.240$

   (e) $\frac{5}{12} \square \frac{3}{10}$

2. Order the following numbers from least to greatest.

   (a) $-5\frac{4}{5}, 5.4, -5\frac{3}{4}, 5.24$

   (b) $-2\frac{3}{4}, -3\frac{3}{5}, -3, 3.\bar{3}, 3.8$

   (c) $6.1, 6.\bar{1}, 6\frac{1}{11}, 6.12, 6$

3. Replace the $\square$ with <, >, or = to make a true statement.

 (a) $-3\frac{1}{4}\square-3.25$

 (b) $\frac{5}{4}\square1.\overline{3}$

 (c) $-3.\overline{4}\square-3\frac{4}{9}$

 (d) $0.\overline{21}\square0.\overline{213}$

 (e) $-\frac{2}{5}\square-0.\overline{42}$

Solutions

1. **(a)** $-6.7>-6.74$ because $-6.70>-6.74$

 (b) $8.63>8.595$ because $8.630>8.595$

 (c) $-\frac{9}{11}<-\frac{7}{11}$ because $-9<-7$

 (d) $-0.24=-0.240$

 (e) $\frac{5}{12}>\frac{3}{10}$ because $\frac{25}{60}>\frac{18}{60}$

2. **(a)** Change to decimals→ −5.80, 5.40, −5.75, 5.24
 Put in order→ −5.80, −5.75, 5.24, 5.40
 Final answer→ $-5\frac{4}{5}$, $-5\frac{3}{4}$, 5.24, 5.4

 (b) Change to decimals→ −2.75, −3.60, −3.00, 3.33 (rounded), 3.80
 Put in order→ −3.60, −3.00, −2.75, 3.33, 3.80
 Final answer→ $-3\frac{3}{5}$, −3, $-2\frac{3}{4}$, $3.\overline{3}$, 3.8

 (c) Change to decimals→ 6.100, 6.111 (rounded), 6.091 (rounded), 6.120, 6.000
 Put in order→ 6.000, 6.091, 6.100, 6.111, 6.120
 Final answer→ 6, $6\frac{1}{11}$, 6.1, $6.\overline{1}$, 6.12

3. **(a)** $-3\frac{1}{4}=-3.25$

 (b) $\frac{5}{4}<1.\overline{3}$ because $1.25<1.33$ (rounded)

 (c) $-3.\overline{4}=-3\frac{4}{9}$

 (d) $0.\overline{21}<0.\overline{213}$ because 0.212 (rounded) < 0.213 (rounded)

 (e) $-\frac{2}{5}>-0.\overline{42}$ because $-0.40>-0.42$ (rounded)

Understanding Absolute Value of Rational Numbers

(CCSS.Math.Content.6.NS.C.6A, CCSS.Math.Content.6.NS.C.7.C, CCSS.Math.Content.6.NS.C.7.D)

The **absolute value** of a rational number is its distance from zero on the number line. The absolute value is indicated by two vertical bars, one on each side of the number. These vertical bars are **absolute value bars.** You read $|-6|$ as "the absolute value of –6." And you read $|6|$ as "the absolute value of 6." As shown below, $|-6| = |6| = 6$ because each is 6 units from zero on the number line.

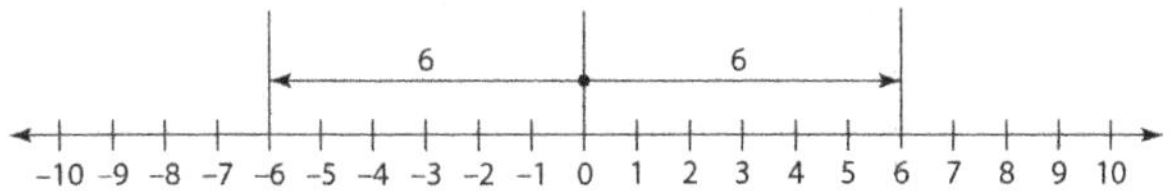

Tip: A number and its opposite have the same absolute value.

Distance *always* has a nonnegative value, meaning it's either positive or zero. So, absolute value is always *nonnegative*. And, for any *nonzero* number, the absolute value is positive. This is true whether the nonzero number is positive or negative.

Here are examples.

$$|-15| = 15,\ \left|\frac{3}{4}\right| = \frac{3}{4},\ |0| = 0,\ |-4.5| = 4.5,\ |-100| = 100,\ |100| = 100$$

As you likely noticed, the absolute values of the numbers in the examples are the values of the numbers with no signs attached. This strategy works for a number whose numerical value you know, but do not use it when you don't know the numerical value of the number. The absolute value of an unknown number n could be n or it could be n's opposite. A good rule of thumb is you can determine the numerical value of the absolute value of a number *only* if you can mark the number's exact location on a number line.

In the previous section, you determined 20 is greater than –100. Now consider their absolute values. Which is greater: $|-100|$ or $|20|$?

To compare two absolute value expressions, find the absolute values first, then compare the resulting nonnegative numbers as shown here.

$|-100| = 100$ and $|20| = 20$, so $|-100|$ is greater than $|20|$ because $100 > 20$.

Think: *"On the number line, –100 is farther from zero than 20, so the absolute value of –100 is greater than the absolute value of 20."*

The order of negative numbers is opposite the order of their absolute values. For instance, –100, –70, and –10 are in increasing order. But their absolute values in increasing order are 10, 70, and 100.

Try These

1. Find the absolute value of the given number.

   (a) 25
   (b) –25
   (c) –13.843
   (d) $7\frac{3}{4}$
   (e) $-7\frac{3}{4}$

2. Replace the $\square$ with <, >, or = to make a true statement.

   (a) $|-6.7| \square |-6.74|$
   (b) $|8.63| \square |8.595|$
   (c) $\left|-\frac{9}{11}\right| \square \left|-\frac{7}{11}\right|$
   (d) $|-95| \square |50|$
   (e) $\left|\frac{5}{12}\right| \square \left|-\frac{3}{10}\right|$

3. Given $-5\frac{3}{5}$, -15, $-5\frac{3}{4}$, -18,

   (a) Order the numbers from least to greatest.
   (b) Order the absolute values of the numbers from least to greatest.
   (c) Are the orderings in parts (a) and (b) different? Why or why not?

Solutions

1. **(a)** 25
   (b) 25
   (c) 13.843
   (d) $7\frac{3}{4}$
   (e) $7\frac{3}{4}$
2. **(a)** $|-6.7| < |-6.74|$ because $6.7 < 6.74$
   (b) $|8.63| > |8.595|$ because $8.630 > 8.595$
   (c) $\left|-\frac{9}{11}\right| > \left|-\frac{7}{11}\right|$ because $\frac{9}{11} > \frac{7}{11}$
   (d) $|-95| > |50|$ because $95 > 50$
   (e) $\left|\frac{5}{12}\right| > \left|-\frac{3}{10}\right|$ because $\frac{25}{60} > \frac{18}{60}$
3. **(a)** $-18, -15, -5\frac{3}{4}, -5\frac{3}{5}$
   (b) $5\frac{3}{5}, 5\frac{3}{4}, 15, 18$
   (c) Yes, the order of negative numbers is opposite the order of their absolute values.

Using Rational Numbers in the Real World

(CCSS.Math.Content.6.NS.C.5, CCSS.Math.Content.6.NS.C.7.B, CCSS.Math.Content.6.NS.C.7.C, CCSS.Math.Content.6.NS.C.7.D)

Rational numbers are used every day in the real world. They are used in science, business, meteorology, oceanography, sports, and various other areas. Interpreting and explaining rational numbers in real-world contexts is an important skill for daily living.

Every nonzero rational number has two components. It has **magnitude** (or absolute value) that tells you "size," and it has a **sign** that tells "direction" as positive or negative. In real-world contexts, "positive or negative" can be interpreted as "right or left," "up or down," "increase or decrease," "above or below," "gain or loss," and so forth.

Zero has a magnitude of 0, but it does not have a sign. Although zero does not have a sign, the signs of the other rational numbers are determined by where they are relative to zero. So it's important to know the meaning of zero for real-world situations.

For instance, banks allow customers to open **checking accounts.** Before any money is put into an account, the account has zero dollars in it. For this situation, the meaning of zero is "no money in the account." The customer can make **transactions.** He or she can make a **deposit** of money into the account or make a **withdrawal** of money from the account. The bank also can make transactions. The bank can put money in by doing a **credit** to the account or take money out by doing a **debit.** The bank can charge **fees** as well. The amount of money in the account at any given time is the account **balance.**

From the customer's viewpoint, *deposits* and *credits* are positive transactions because they increase the account balance. *Withdrawals, debits,* and *fees* are negative transactions because they decrease the balance. Using rational numbers to represent transactions is a good way to keep up with the account balance. Here are examples of using rational numbers to represent money transactions.

Transaction	Represented as
Deposit of $250	$250
Withdrawal of $75	-$75
Debit of $23.90	-$23.90
Fee of $4	-$4
Credit of $8.95	$8.95

Use absolute value to make comparisons of the sizes of transactions. For instance, a debit of $45 decreases the account more than a debit of $20 because |-$45| > |-$20|. Together, a withdrawal of $100 and a deposit of $100 result in no change in the balance because |$100| = |-$100|.

Altitude is another real-world context for rational numbers. The zero point is sea level. Altitudes above sea level are positive, and those below sea level are negative. An altitude of 2,500 feet is higher than an altitude of 1,000 feet. An altitude of –200 feet is lower than an altitude of –50 feet. Here are examples of using rational numbers to represent altitudes.

Altitude	Represented as
Mount McKinley is 20,322 feet above sea level.	20,322 feet
New Orleans, Louisiana, is 7 feet below sea level.	-7 feet
Bombay Beach, California, is 226 feet below sea level.	-226 feet
Mobile, Alabama, is 218 feet above sea level.	218 feet

Use absolute value to compare magnitudes of altitudes. The magnitude of an altitude of 218 feet is less than the magnitude of an altitude of –226 feet because |218 feet| < |–226 feet|. So, Mobile, Alabama, is closer to sea level than is Bombay Beach, California.

Temperature is measured in degrees. Temperatures above zero degrees are recorded as positive, and those below zero degrees are recorded as negative. The meaning of the zero point depends on which temperature scale you are using. For the Celsius thermometer, scientists have agreed 0°C designates the freezing (or melting) point of water. On the Fahrenheit thermometer, the zero point is 32 degrees below the freezing point of water. So a temperature of 0°F is *not* equivalent to a temperature of 0°C.

Tip: Read "0°C" as "zero degrees Celsius" and "0°F" as "zero degrees Fahrenheit."

Here is a side-by-side comparison of Celsius and Fahrenheit thermometers.

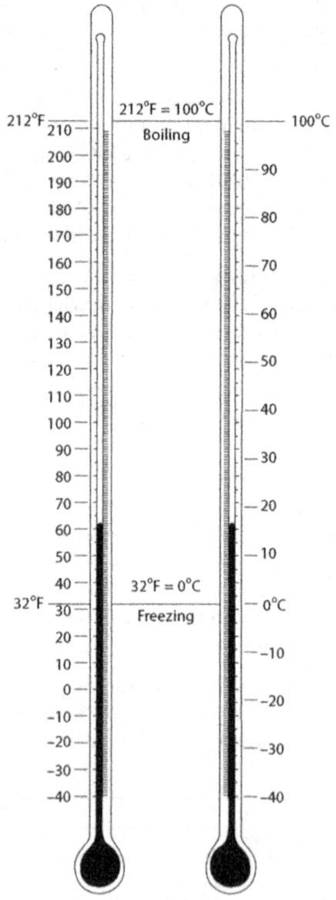

A temperature of 35°C is warmer than a temperature of 20°C. A temperature of –15°C is colder than a temperature of –9°C. Similarly, 90°F is warmer than 72°F and –10°F is colder than –2°F.

Here are examples of using rational numbers to represent temperature readings.

Real-World Benchmark	**Fahrenheit Temperature**	**Represented as**	**Celsius Temperature**	**Represented as**
Freezing/melting point of water	32 degrees above zero	32°F	zero degrees	0°C
Boiling point of water	212 degrees above zero	212°F	100 degrees above zero	100°C
Average winter temperature in Antarctica	29.9 degrees below zero	–29.9°F	34.4 degrees below zero	–34.4°C
Average room temperature	70 degrees above zero	70°F	21 degrees above zero	21°C

A temperature of 21°C is closer to the freezing/melting point of water (0°C) than is a temperature of –34.4°C because |21°C| < |–34.4°C|.

☞ Try These

1. Use rational numbers to represent each of the following bank account transactions.
   (a) A deposit of $1,000
   (b) A debit of $86.54
   (c) A fee of $5
   (d) A withdrawal of $350
   (e) A credit of $25.67

2. Use rational numbers to represent each of the following altitudes.
   (a) 5,280 feet above sea level (Denver, Colorado)
   (b) 427 meters below sea level (Dead Sea, Jordan/Israel)
   (c) 29,029 feet above sea level (Mount Everest, Himalayas)
   (d) 282 feet below sea level (Death Valley, Mojave Desert, California)
   (e) 489 feet above sea level (Austin, Texas)

3. Fahrenheit temperature readings in a northern location in the United States during a 5-day period in winter were –1°F, 10°F, –5°F, 15°F, and –8°F. Order these temperatures from coldest to warmest.

4. Which temperature is closer to the freezing/melting point of water (0°C)?
   (a) –25.4°C or –28.5°C
   (b) 15°C or –18°C

Solutions

1. **(a)** $1,000
   (b) –$86.54
   (c) –$5
   (d) –$350
   (e) $25.67

2. **(a)** 5,280 feet
   (b) –427 meters
   (c) 29,029 feet
   (d) –282 feet
   (e) 489 feet
3. –8°F, –5°F, –1°F, 10°F, 15°F
4. **(a)** –25.4°C is closer because |–25.4°C| < |–28.5°C|.
   (b) 15°C is closer because |15°C| < |–18°C|.

Plotting Points and Finding Distances Between Points in the Coordinate Plane

(CCSS.Math.Content.6.NS.C.6.A, CCSS.Math.Content.6.NS.C.6.B, CCSS.Math.Content.6.NS.C.6.C, CCSS.Math.Content.6.NS.C.8)

If you take two copies of the number line, one horizontal and one vertical, and position them at right angles so that they intersect at the 0 point on each line, you have a **coordinate plane.**

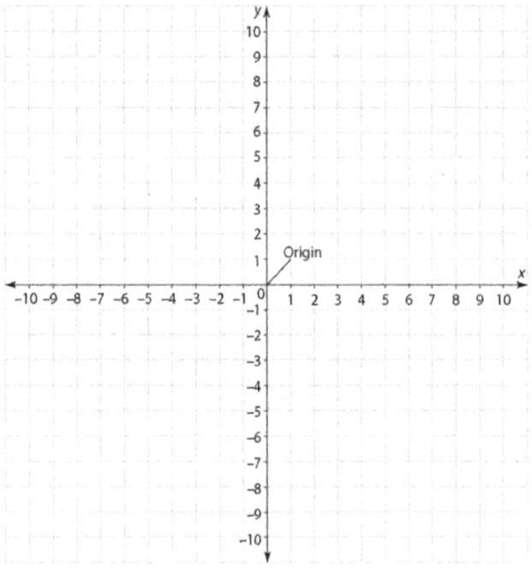

Commonly, the horizontal number line with positive direction to the right is designated the ***x*-axis,** and the vertical number line with positive direction upward is designated the ***y*-axis.** The intersection of the two lines is the **origin.** The two intersecting *x*- and *y*-axes divide the coordinate plane into four sections, called **quadrants.**

The quadrants are numbered *counterclockwise* using Roman numerals as shown here.

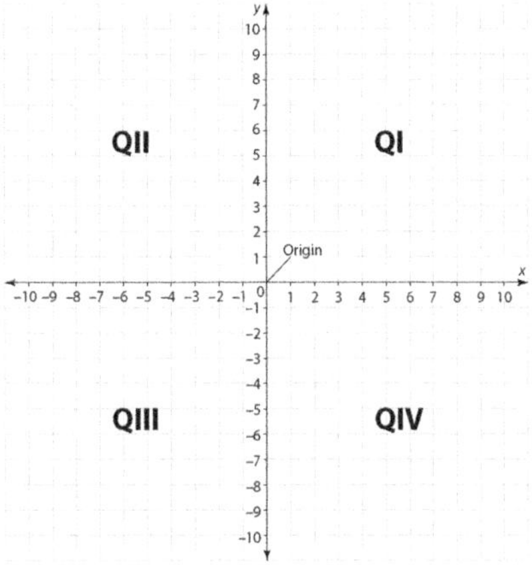

Plotting Points in the Coordinate Plane

In the coordinate plane, you match points with an **ordered pair** of numbers called the **coordinates** of the point. An ordered pair of numbers is written in a definite order so that one number is first and the other is second. The first number is the **first coordinate** (or x-coordinate), and the second number is the **second coordinate** (or y-coordinate). Write ordered pairs in parentheses with the two numbers separated by a comma. The ordered pair (0, 0) designates the origin.

Look at these examples.

(3, 5) is the ordered pair with first coordinate = 3 and second coordinate = 5

(–2, 3) is the ordered pair with first coordinate = –2 and second coordinate = 3

An ordered pair gives you directions on how to graph a point in the coordinate plane, starting from the origin (0, 0). The first coordinate tells you how far to go right or left. For positive numbers, go right; for negative numbers, go left. From that spot, the second coordinate tells you how far to go up or down. For positive numbers, go up; for negative numbers, go down. Then, insert a large dot to mark the location of the point.

Here are examples of how to graph a point.

Graph the ordered pair (3, 5).

Start at (0, 0). Go right 3 units. Then from there go up 5 units. Mark the location.

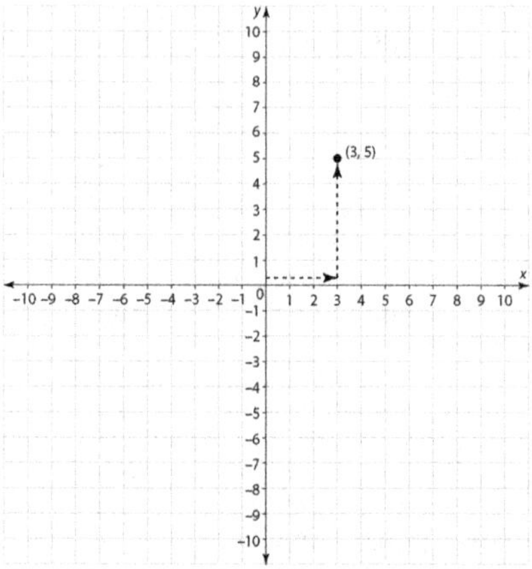

Graph the ordered pair (–2, 3).

Start at (0, 0). Go left 2 units. Then from there go up 3 units. Mark the location.

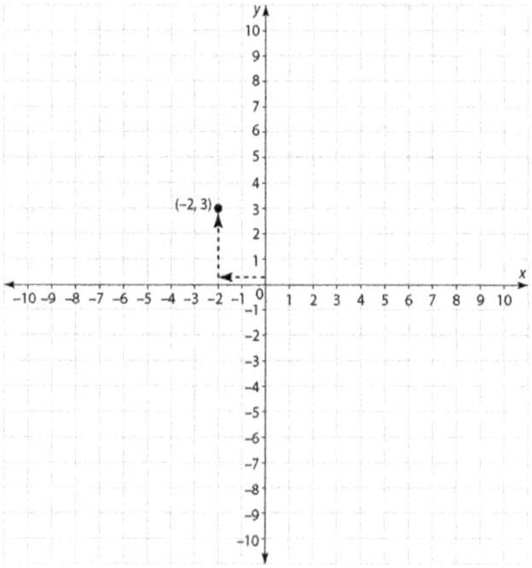

Graph the ordered pair (–4, –5).

Start at (0, 0). Go left 4 units. Then from there go down 5 units. Mark the location.

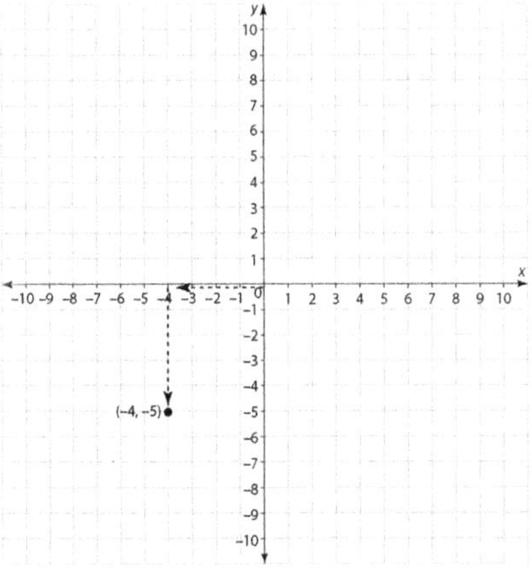

Graph the ordered pair (1, –6).

Start at (0, 0). Go right 1 unit. Then from there go down 6 units. Mark the location.

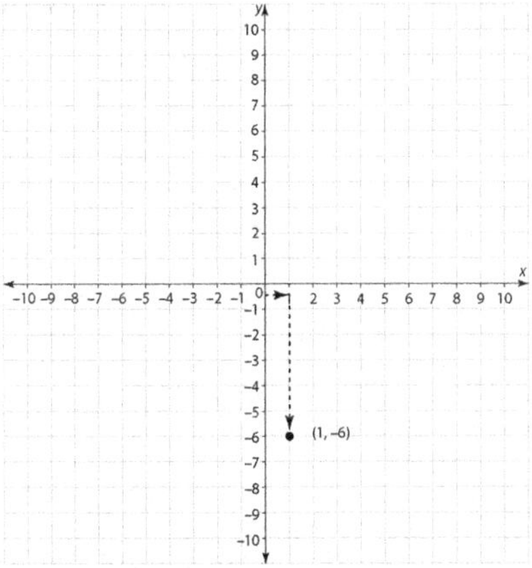

For convenience, use "**x-coordinate**" to mean the first coordinate of an ordered pair and use "**y-coordinate**" to mean the second coordinate. In Quadrant I, both the x-coordinate and the y-coordinate are positive; in Quadrant II, the x-coordinate is negative, and the y-coordinate is positive; in Quadrant III, both the x-coordinate and the y-coordinate are negative; and in Quadrant IV, the x-coordinate is positive, and the y-coordinate is negative.

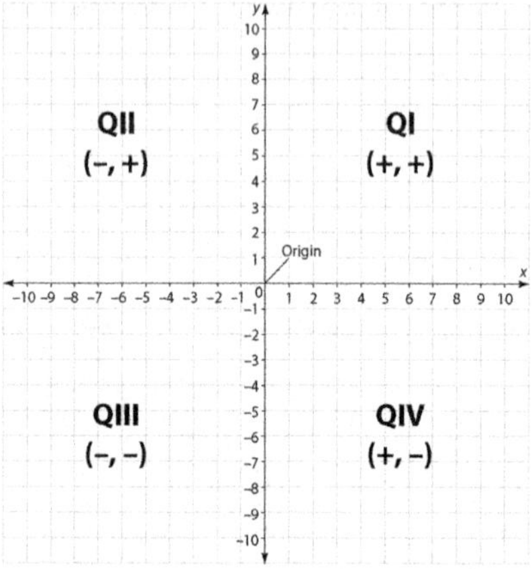

Points that have zero as one or both of the coordinates lie on one or both of the axes. A point whose x-coordinate is 0 lies on the y-axis because the point is located 0 units right or left from the y-axis. A point whose y-coordinate is 0 lies on the x-axis because the point is located 0 units above or below the x-axis. If both coordinates of a point are 0, the point is at the origin. For example, as shown below, the point (0, 4) lies on the y-axis, the point (–5, 0) lies on the x-axis, and the point (0, 0) is at the origin.

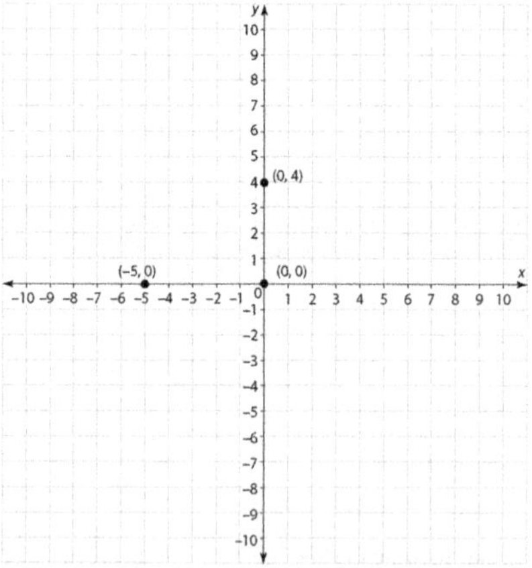

You name a point in a coordinate plane by naming the ordered pair (x, y) that specifies the location of the point. The location of every point in the coordinate plane is given by an ordered pair of numbers. The numbers x and y are the coordinates of the point. Here is an example.

What ordered pair of integers represents the point K?

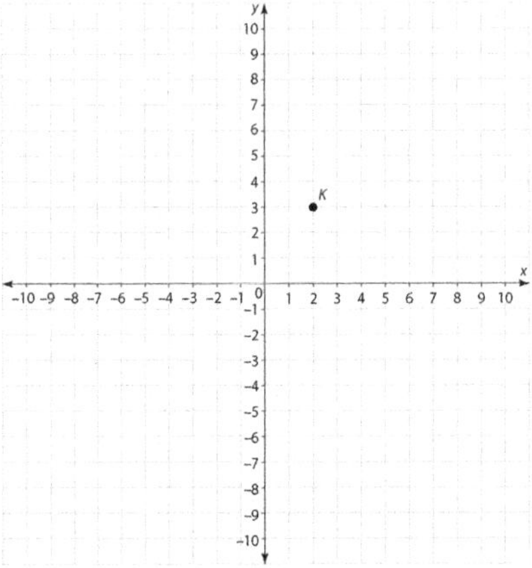

The point K is 2 units to the right and 3 units up from the origin. The ordered pair (2, 3) represents the point K.

Two ordered pairs represent the same point in the coordinate plane if and only if their corresponding coordinates are equal. So $(a, b) = (c, d)$ if and only if $a = c$ and $b = d$. For example, as shown below, (3, 5) and (5, 3) do not represent the same point in the coordinate plane. The two x-coordinates, 3 and 5, are not equal. Also, the two y-coordinates, 5 and 3, are not equal.

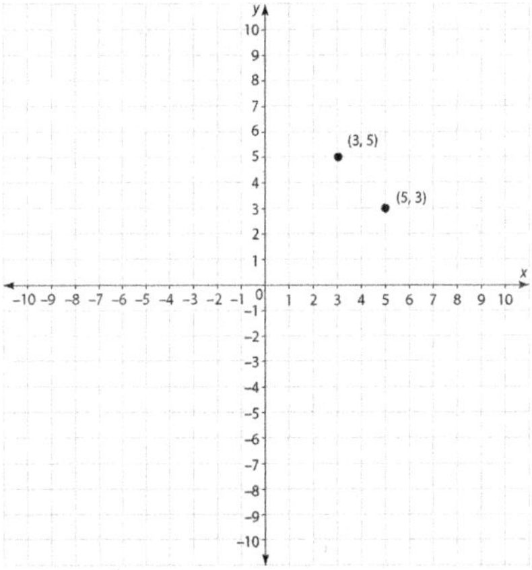

The point $(a, -b)$ is a **reflection across the x-axis** of the point (a, b). Specifically, $(a, -b)$ is the **image** of (a, b) when (a, b) is reflected across the x-axis. Look at this example.

Plot the points (3, 5) and (3, –5) in the same coordinate plane.

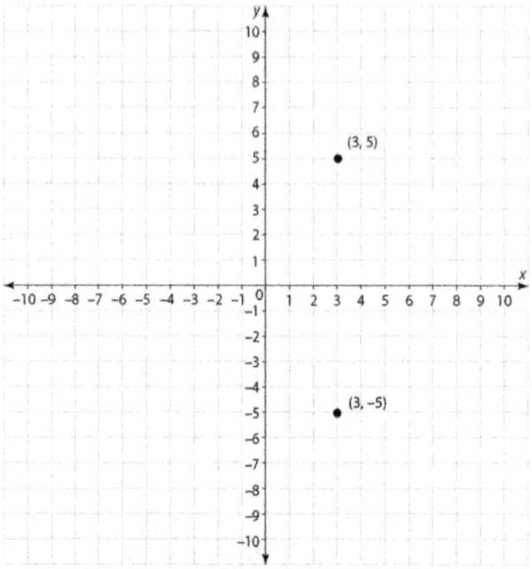

The points have the same x-coordinate, so they are the same distance to the right of the y-axis. Their y-coordinates are opposites, so they are the same distance from the x-axis, but on opposite sides. The point (3, –5) is the image of (3, 5) when (3, 5) is reflected across the x-axis.

The point $(-a, b)$ is a **reflection across the y-axis** of the point (a, b). Specifically, $(-a, b)$ is the **image** of (a, b) when (a, b) is reflected across the y-axis. Look at this example.

Plot the points (3, 5) and (–3, 5) in the same coordinate plane.

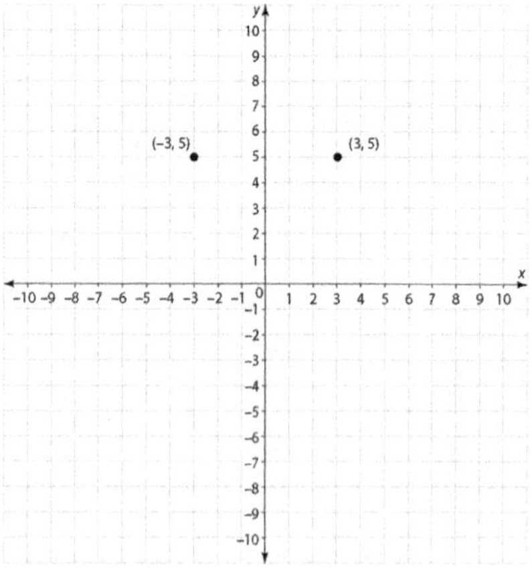

The points have the same y-coordinate, so they are the same distance above the x-axis. Their x-coordinates are opposites, so they are the same distance from the y-axis, but on opposite sides. The point (–3, 5) is the image of (3, 5) when (3, 5) is reflected across the y-axis.

The point $(-a, -b)$ is a **reflection across the origin** of the point (a, b). Specifically, $(-a, -b)$ is the **image** of (a, b) when (a, b) is reflected across the origin. Look at this example.

Plot the points (3, 5) and (–3, –5) in the same coordinate plane.

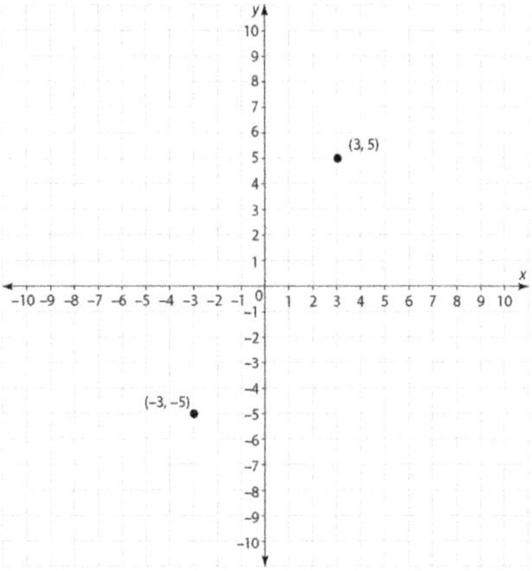

The x-coordinates of the points are opposites, so they are the same distance from the y-axis, but on opposite sides. Their y-coordinates are opposites, so they are the same distance from the x-axis, but on opposite sides. The point (–3, –5) is the image of the point (3, 5) when (3, 5) is reflected across the origin.

Try These

1. Name the quadrant in which the ordered pair is located.

   (a) (14, –16)
   (b) (20, 35)
   (c) (–15, –40)
   (d) (–8, 12)

2. Fill in the blank.

   (a) Every point on the x-axis has a y-coordinate equal to __________.
   (b) Every point on the y-axis has an x-coordinate equal to__________.
   (c) The point (0, 10) lies on the __________ (x-axis, y-axis).
   (d) The point (–5, 0) lies on the __________ (x-axis, y-axis).
   (e) The ordered pairs (2, 7) and (7, 2) are __________ (equal, not equal).

3. Match the points on the graph with the given ordered pairs.

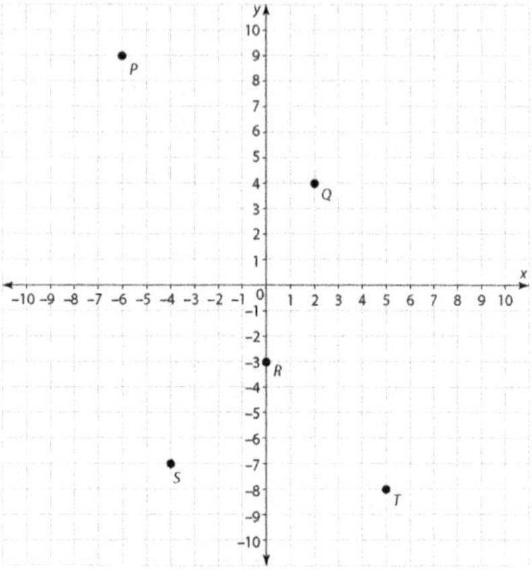

(a) (2, 4)
(b) (5, –8)
(c) (0, –3)
(d) (–4, –7)
(e) (–6, 9)

4. Given triangle *ABC* shown, fill in the coordinates in the chart below.

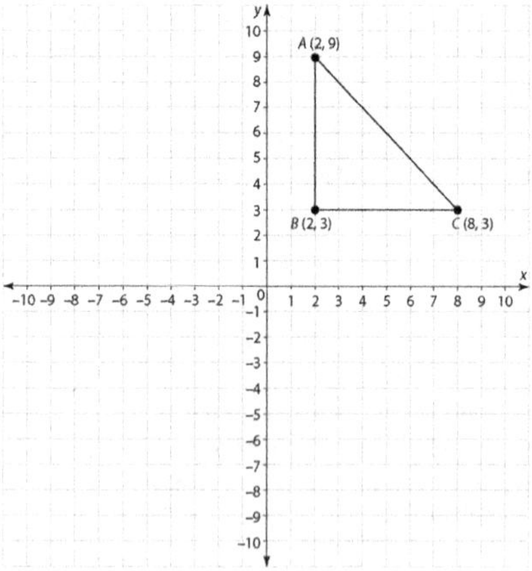

Reflection	Image of *A*	Image of *B*	Image of *C*
Over *x*-axis			
Over *y*-axis			
Over origin			

Solutions

1. **(a)** IV
   (b) I
   (c) III
   (d) II
2. **(a)** zero
   (b) zero
   (c) y-axis
   (d) x-axis
   (e) not equal
3. **(a)** Q
   (b) T
   (c) R
   (d) S
   (e) P
4.

Reflection	Image of *A*	Image of *B*	Image of *C*
Over *x*-axis	(2, –9)	(2, –3)	(8, –3)
Over *y*-axis	(–2, 9)	(–2, 3)	(–8, 3)
Over origin	(–2, –9)	(–2, –3)	(–8, –3)

Finding Distances Between Points in the Coordinate Plane

Two points that have the same x-coordinate lie on the same vertical line. For instance, (3, 5) and (3, 1) lie on the same vertical line.

Similarly, two points that have the same y-coordinate lie on the same horizontal line. For instance, (1, –6) and (–7, –6) lie on the same horizontal line.

To find the distance between two points that have the same x-coordinate and lie in the same quadrant, *subtract* the smaller of the absolute values of the y-coordinates of the two points from the larger absolute value. Here is an example.

> Find the distance between the points (3, 5) and (3, 1) in the coordinate plane. The x-coordinates of the two points are the same. The y-coordinates have the same sign. A graph shows the two points lie in the same quadrant.

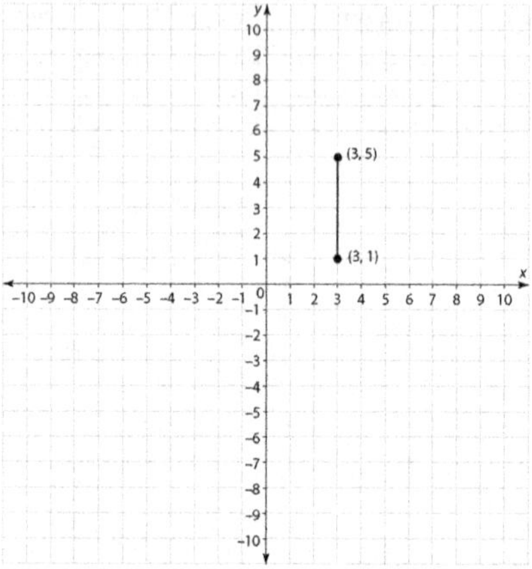

The distance between the points is the length of the vertical line segment connecting them. To find the length of this segment, subtract the smaller of the absolute values of the *y*-coordinates from the larger absolute value.

$|5| - |1| = 5 - 1 = 4$, so the distance between (3, 5) and (3, 1) is 4 units. *Check:* By counting, the distance also is 4 units. ✓

To find the distance between two points that have the same *x*-coordinate and lie in different quadrants, *add* the absolute values of the *y*-coordinates of the two points. Here is an example.

> Find the distance between the points (–2, 3) and (–2, –4) in the coordinate plane. The *x*-coordinates of the two points are the same. The *y*-coordinates have opposite signs. A graph shows the two points lie in different quadrants.

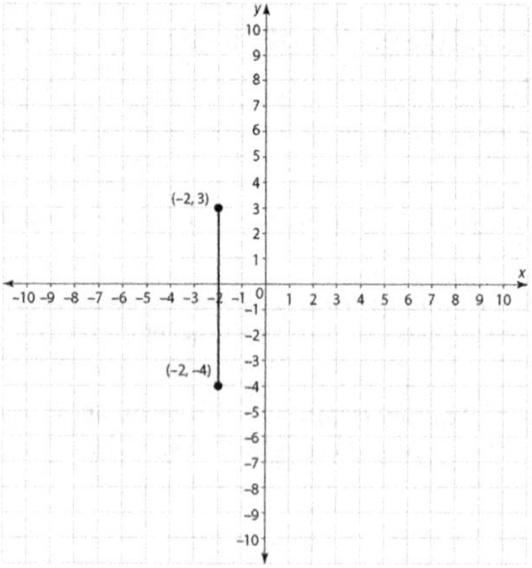

The distance between the points is the length of the vertical line segment connecting them. To find the length of this segment, add the absolute values of the y-coordinates.

$|3| + |-4| = 3 + 4 = 7$, so the distance between (–2, 3) and (–2, –4) is 7 units. *Check:* By counting, the distance also is 7 units. ✓

To find the distance between two points that have the same y-coordinate and lie in the same quadrant, *subtract* the smaller of the absolute values of the x-coordinates of the two points from the larger absolute value. Here is an example.

Find the distance between the points (–4, –5) and (–9, –5) in the coordinate plane. The y-coordinates of the two points are the same. The x-coordinates have the same sign. A graph shows the two points lie in the same quadrant.

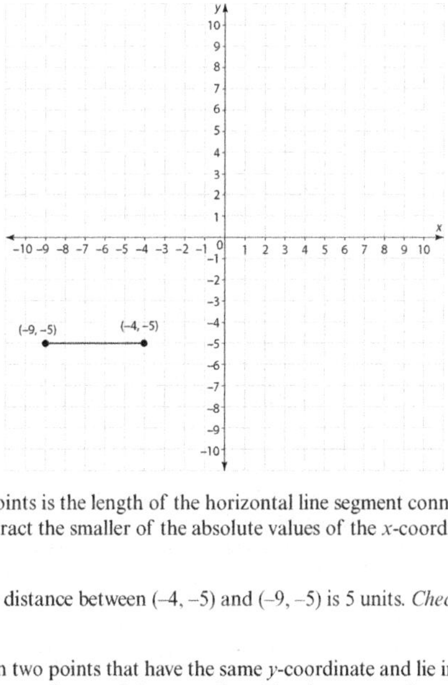

The distance between the points is the length of the horizontal line segment connecting them. To find the length of this segment, subtract the smaller of the absolute values of the x-coordinates from the larger absolute value.

$|-9| - |-4| = 9 - 4 = 5$, so the distance between (–4, –5) and (–9, –5) is 5 units. *Check:* By counting, the distance also is 5 units. ✓

To find the distance between two points that have the same y-coordinate and lie in different quadrants, *add* the absolute values of the x-coordinates of the two points. Here is an example.

Find the distance between the points (1, –6) and (–7, –6) in the coordinate plane. The y-coordinates of the two points are the same. The x-coordinates have opposite signs. A graph shows the two points lie in different quadrants.

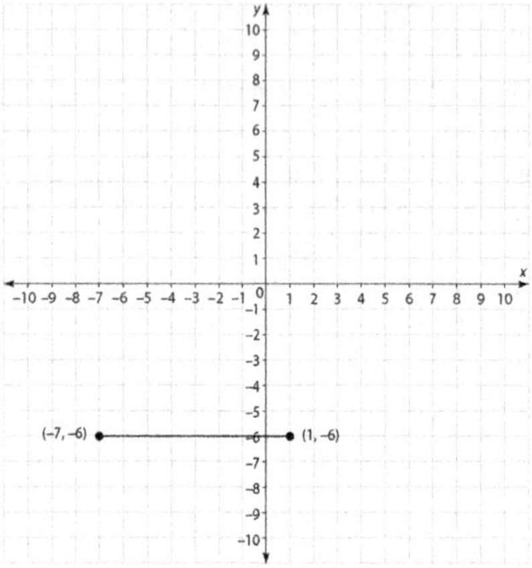

The distance between the points is the length of the horizontal line segment connecting them. To find the length of this segment, add the absolute values of the x-coordinates.

$|-7| + |1| = 7 + 1 = 8$, so the distance between (1, –6) and (–7, –6) is 8 units. *Check:* By counting, the distance also is 8 units. ✓

Here is a summary of finding vertical and horizontal distances between two points in the coordinate plane.

To find the distance between two points that lie on the same vertical line, *subtract* the smaller of the absolute values of their y-coordinates from the larger absolute value if the y-coordinates have the same sign, but *add* the absolute values of their y-coordinates if the y-coordinates have different signs.

Similarly, to find the distance between two points that lie on the same horizontal line, *subtract* the smaller of the absolute values of their x-coordinates from the larger absolute value if the x-coordinates have the same sign, but *add* the absolute values of their x-coordinates if the x-coordinates have different signs.

☞Try These

1. Fill in the blank.

   (a) Two points with the same __________ (x-coordinate, y-coordinate) lie on the same vertical line.

   (b) Two points with the same __________ (x-coordinate, y-coordinate) lie on the same horizontal line.

   (c) The points (–30, 25) and (–30, –18) lie on the same __________ (horizontal, vertical) line.

   (d) The points (14, 20) and (100, 20) lie on the same __________ (horizontal, vertical) line.

2. Find the distance between the two points.
   (a) (8, –15) and (8, –3)
   (b) (75, –12) and (–50, –12)
   (c) (14, 20) and (100, 20)
   (d) (–30, 25) and (–30, –18)

3. Find the perimeter of rectangle $ABCD$.

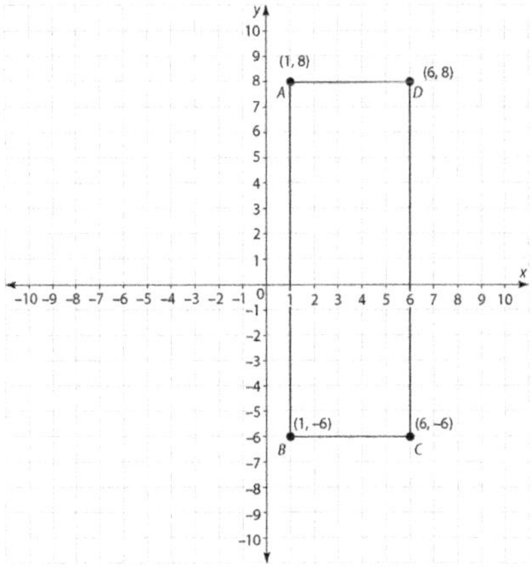

Solutions

1. **(a)** x-coordinate
   (b) y-coordinate
   (c) vertical
   (d) horizontal
2. **(a)** $|-15| - |-3| = 15 - 3 = 12$
   (b) $|75| + |-50| = 75 + 50 = 125$
   (c) $|100| - |-14| = 100 - 14 = 86$
   (d) $|25| + |-18| = 25 + 18 = 43$
3. The perimeter of rectangle $ABCD = AB + BC + CD + AD$.
   $AB = CD = |8| + |-6| = 8 + 6 = 14$; $BC = AD = |6| - |1| = 6 - 1 = 5$;
   perimeter = 14 + 14 + 5 + 5 = 38 units

3. Expressions and Equations

In this chapter, you will learn about variable expressions and equations. You will extend your understanding of arithmetic to include letters to represent unknown quantities. You will evaluate mathematical expressions and formulas, solve real-world and mathematical equations, and use inequalities to represent constraints in real-life and mathematical situations. In addition, you will use variables to represent two quantities in real-world and mathematical problems that change in relationship to one another.

Understanding and Evaluating Exponential Expressions Involving Whole-Number Exponents

(CCSS.Math.Content.6.EE.A.1)

Suppose you have the problem 2 + 2 + 2 + 2 + 2. This problem is an example of repeated addition. **Repeated addition** means using the *same* number as an *addend* many times. A shorthand way to express 2 + 2 + 2 + 2 + 2 is 5 × 2.The expression 5 × 2 means "use 2 as an addend 5 times." ***Tip:*** Because you've memorized your multiplication tables, you know the product is 10 without actually adding up five 2s.

Now suppose you have the problem 2 × 2 × 2 × 2 × 2. This problem is an example of repeated multiplication. **Repeated multiplication** means using the *same* number as a *factor* many times. You use exponents to express repeated multiplication.

An **exponent** is a small raised number written to the upper right of a quantity, which is called the **base** for the exponent. For the product 2 × 2 × 2 × 2 × 2, write 2^5. The expression 2^5 means "use 2 as a factor 5 times." The number 2 is the base, and the small 5 to the upper right of 2 is the exponent.

$$2^5 \begin{array}{l}\leftarrow \text{Exponent} \\ \leftarrow \text{Base}\end{array}$$

Tip: Exponents are written as superscripts, so they are smaller than the other numbers in a problem. Write an exponent slightly raised and immediately to the right of the number that is its base. Do this carefully so that, for example, 2^5 is not mistaken for 25.

An exponent, n, that is a nonzero whole number tells how many times the base, b, is used as a factor. That is, $b^n = \underbrace{b \times b \times \cdots b}_{n \text{ times}}$.

Expressions like 2^5 are **exponential expressions.** To **evaluate** an exponential expression, do to the base what the exponent tells you to do. Evaluating exponential expressions is **exponentiation.** Here are examples.

$2^5 = 2 \times 2 \times 2 \times 2 \times 2 = 32$ ***Tip:*** Notice 2^5 is *not* 2 × 5. 2^5 is 32; 2 × 5 is 10.

$3^4 = 3 \times 3 \times 3 \times 3 = 81$

$10^6 = 10 \times 10 \times 10 \times 10 \times 10 \times 10 = 1{,}000{,}000$

$6^2 = 6 \times 6 = 36$

$5^3 = 5 \times 5 \times 5 = 125$

$7^1 = 7$

Read 2^5 as "two to the fifth power." Read 3^4 as "three to the fourth power." Read 10^6 as "ten to the sixth power," and so on. When 2 is the exponent, as in 6^2, you say "six squared," and when 3 is the exponent, as in 5^3, you say "five cubed." When no exponent is written on a number, the exponent is understood to be 1. So 7 is 7^1, read "7 to the first power."

When the exponent is a nonzero whole number, the base can be any rational number. Enclosing fractions and decimals in parentheses and using parentheses to indicate multiplication can make the problem easier to read. Here are examples.

$(0.1)^5 = (0.1)(0.1)(0.1)(0.1)(0.1) = 0.00001$

$$\left(\frac{3}{4}\right)^3 = \left(\frac{3}{4}\right)\left(\frac{3}{4}\right)\left(\frac{3}{4}\right) = \frac{27}{64}$$

$(2.5)^2 = (2.5)(2.5) = 6.25$

$$\left(3\frac{1}{2}\right)^2 = \left(\frac{7}{2}\right)^2 = \left(\frac{7}{2}\right)\left(\frac{7}{2}\right) = \frac{49}{4} = 12\frac{1}{4}$$

Tip: In an exponential expression, which number is the exponent and which is the base makes a difference in the value of the expression. Unlike multiplication, where $5 \times 2 = 2 \times 5 = 10$, exponential expressions do NOT have a commutative property. So $2^5 \neq 5^2$; $2^5 = 2 \times 2 \times 2 \times 2 \times 2 = 32$ and $5^2 = 5 \times 5 = 25$.

☞ Try These

1. Fill in the blank(s).

   (a) In the exponential expression 10^4, __________ is the exponent and __________ is the base.

   (b) An exponent that is a nonzero whole number tells how many times the base is used as a __________.

   (c) The expression 5^4 is read "five to the fourth __________."

   (d) When no exponent is written on a number, the exponent is understood to be __________.

2. Write the indicated product as an exponential expression.

   (a) $5 \times 5 \times 5 \times 5$

   (b) $(1.2)(1.2)(1.2)$

   (c) $\left(\frac{1}{2}\right)\left(\frac{1}{2}\right)\left(\frac{1}{2}\right)\left(\frac{1}{2}\right)\left(\frac{1}{2}\right)$

   (d) 200×200

3. Write an exponential expression for the verbal phrase.
   (a) seven to the fifth power
   (b) five to the fourth power
   (c) nine squared
   (d) two cubed

4. Evaluate the exponential expression.
   (a) 5^4
   (b) $(1.2)^3$
   (c) $\left(\frac{1}{2}\right)^5$
   (d) 200^2

Solutions

1. **(a)** 4; 10
   (b) factor
   (c) power
   (d) 1
2. **(a)** 5^4
   (b) $(1.2)^3$

Tip: Remember to use parentheses when writing exponential expressions that contain decimals or fractions.

   (c) $\left(\frac{1}{2}\right)^5$
   (d) 200^2
3. **(a)** 7^5
   (b) 5^4
   (c) 9^2
   (d) 2^3
4. **(a)** $5 \times 5 \times 5 \times 5 = 625$
   (b) $(1.2)(1.2)(1.2) = 1.728$
   (c) $\left(\frac{1}{2}\right)\left(\frac{1}{2}\right)\left(\frac{1}{2}\right)\left(\frac{1}{2}\right)\left(\frac{1}{2}\right) = \frac{1}{32}$
   (d) $200 \times 200 = 40{,}000$

Evaluating Numerical Expressions

(CCSS.Math.Content.6.EE.A.1)

A **numerical expression** is a number or two or more numbers with indicated operations, such as addition, subtraction, multiplication, or division that evaluates to a number: $4 + 8$, $32 - 20$, $\frac{15}{3}$, 2×5, $7(2 + 3)$, or 6^2.

When more than one operation is involved in a numerical expression, you must follow the **order of operations** to evaluate the expression. Proceed in a step-by-step manner.

1. Compute inside **Parentheses** (or other grouping symbols).
2. Do **Exponentiation** (that is, evaluate exponential expressions).
3. **Multiply** and **Divide** in the order in which they occur from left to right.
4. **Add** and **Subtract** in the order in which they occur from left to right.

Tip: Multiplication does not always have to be done before division, or addition before subtraction. You multiply and divide in the order the operations occur in the problem from left to right . Similarly, you add and subtract in the order they occur in the problem.

Of course, not every problem will require you to use all four steps of the process or to perform all operations. Just do the operations given in the problem in the proper order. Make sure you don't skip a step you are supposed to do before another step.

Here are examples.

Evaluate $2 + 4 \times 5 - 6 \div 2$.

There are no parentheses or exponents, so multiply and divide, then add and subtract.

$$2 + 4 \times 5 - 6 \div 2 = 2 + 20 - 3 = 19$$

Evaluate $6^2 \div 12 + 5 \times 2$.

There are no parentheses, so do exponentiation, multiply and divide, then add.

$$6^2 \div 12 + 5 \times 2 = 36 \div 12 + 5 \times 2 = 3 + 10 = 13$$

Evaluate $25 - 3(4 + 2)$.

Compute inside parentheses. There are no exponents, so multiply, then add and subtract.

$$25 - 3(4 + 2) = 25 - 3(6) = 25 - 18 = 7$$

Tip: Grouping symbols, such as parentheses, keep things together that belong together. Always do the operations inside grouping symbols first—especially when you have addition and/or subtraction inside the grouping symbols. Otherwise, you might get an incorrect result. When you no longer need the grouping symbols, omit them.

Evaluate $12 \times 5 + 3^4 \div (4 + 5) - 10$.

Compute inside parentheses, do exponentiation, multiply and divide, then add and subtract.

$$12 \times 5 + 3^4 \div (4 + 5) - 10 = 12 \times 5 + 3^4 \div (9) - 10 = 12 \times 5 + 81 \div 9 - 10 = 60 + 9 - 10 = 59$$

Tip: Here is a sentence to help you remember the order of operations: Please Excuse My Dear Aunt Sally—abbreviated PE(MD)(AS). The first letter of each word gives the order of operations: Parentheses, Exponentiation, Multiply and Divide, Add and Subtract. The parentheses around MD and AS are to remind you to multiply and divide or add and subtract in the order they occur in the problem.

Try These

1. Evaluate $20 - 3 \times 5 - 8 \div 2$.
2. Evaluate $8^2 \div 16 + 7 \times 3$.
3. Evaluate $47 + 3(10 - 2)$.
4. Evaluate $90 - 5 \times 3^2 + 42 \div (5 + 2)$.
5. Evaluate $\frac{21}{3} + 2^3(7-4) + 5^2 - 13$.

Solutions

1. There are no parentheses or exponents, so multiply and divide, then subtract.

$$20 - 3 \times 5 - 8 \div 2 = 20 - 15 - 4 = 1$$

2. There are no parentheses, so do exponentiation, multiply and divide, then add.

$$8^2 \div 16 + 7 \times 3 = 64 \div 16 + 7 \times 3 = 4 + 21 = 25$$

3. Compute inside parentheses. There are no exponents, so multiply, then add.

$$47 + 3(10 - 2) = 47 + 3(8) = 47 + 24 = 71$$

4. Compute inside parentheses, do exponentiation, multiply and divide, then add and subtract.

$$90 - 5 \times 3^2 + 42 \div (5 + 2) = 90 - 5 \times 3^2 + 42 \div (7) = 90 - 5 \times 9 + 42 \div 7 = 90 - 45 + 6 = 51$$

5. Compute inside parentheses, do exponentiation, multiply and divide, then add and subtract.

$$\frac{21}{3} + 2^3(7-4) + 5^2 - 13 = \frac{21}{3} + 2^3(3) + 5^2 - 13 = \frac{21}{3} + 8(3) + 25 - 13 = 7 + 24 + 25 - 13 = 43$$

Understanding Variables and Expressions in Which Letters Stand for Numbers

(CCSS.Math.Content.6.EE.A.2.B, CCSS.Math.Content.6.EE.B.6)

A **mathematical expression** uses a symbol or a combination of symbols to represent a number. The symbols can be arithmetic numbers, letters, and operation symbols. The arithmetic numbers are **constants.** You know their values. The **letters** such as x or n, for example, stand for numbers. Without further information, you do not know the value of x or n. The **operation symbols** indicate calculations to be performed.

Tip: When the meaning is clear, refer to mathematical expressions as "expressions."

An expression can be a numerical expression, as in the previous section. The expression 3(5 + 7) is a product of two factors, 3 and (5 + 7). The factor (5 + 7) is the sum of 5 and 7. The expression 3(5 + 7) is the product of 3 and the sum of 5 + 7. The value of 3(5 + 7) is 3(12), or 36. An expression can be a single number, even 0.

A letter in an expression represents a **variable.** The letter is the variable's "name." You refer to variable x simply as "x," its letter representation. The letter is a placeholder for a number or, in some cases, a range of numbers. For instance, you can replace the number 6 with the letter x in the equality statement 6 + 4 = 10 to get $x + 4 = 10$. In this expression, the letter x is holding a place open for the number 6.

The expression $x + 4$ represents a number. That number is 10 when x is 6. Use different letters to stand for different numbers in an expression. For instance, you can replace the number 6 with the letter x and the number 4 with the letter y in the equality statement 6 + 4 = 10 to get $x + y = 10$.

In problem situations, you use variables to represent unknown numbers. Upper- or lowercase letters like x, y, z, A, B, or C name the variables. The letters are standing for specific numbers, but the values are unknown. Your task is to figure out which numbers the letters stand for. The rules of arithmetic apply to the letters the same as they do to all numbers.

A **term** is a number, a variable, or an indicated product or quotient of numbers and variables. In an expression, terms are separated by plus or minus signs. Here are examples.

The expression $x + 4$ has two terms. The terms are x and 4.

The expression $20y$ has one term.

The expression 3(5 + 7) has one term. ***Tip:*** The term 3(5 + 7) is an indicated product, so it counts as only one term in the expression.

The expression $5x - 2y + 1$ has three terms. The terms are $5x$, $2y$, and 1.

The expression $\frac{21}{3} + 2^3(7 - 4) + 5^2 - 13$ has four terms. The terms are $\frac{21}{3}$, $2^3(7 - 4)$, 5^2, and 13.

Try These

1. Fill in the blank.

   (a) An expression uses symbols to represent a __________.
   (b) The arithmetic numbers in an expression are __________.
   (c) The letters in an expression are placeholders for __________.
   (d) A variable is represented by a __________ in an expression.
   (e) In an expression, __________ are separated by plus or minus signs.

2. State the number of terms in the given expression.
   (a) $10(1 + 5x)$
   (b) $2(a + 1) - 3 + 4a$
   (c) $x + x + x + y + y$
   (d) $\frac{4x}{5} + 3(7x) + 6^2 - 12$
   (e) $12x + 48$

Solutions

1. **(a)** number
   (b) constants
   (c) numbers
   (d) letter
   (e) terms
2. **(a)** one
   (b) three
   (c) five
   (d) four
   (e) two

Writing and Reading Expressions in Which Letters Stand for Numbers

(CCSS.Math.Content.6.EE.A.2.A, CCSS.Math.Content.6.EE.A.2.B, CCSS.Math.Content.6.EE.B.6)

Writing and reading expressions that record operations with numbers and with letters standing for numbers is an important problem-solving skill. In this section, you will express addition, subtraction, multiplication, and division in algebraic form. **Algebraic form** is the form you use when you have variables in expressions. You also will read expressions using correct terminology for operation symbols.

Tip: When solving problems, you might find it helpful to use the first letter of a key word to represent an unknown variable.

Writing and Reading Addition Expressions

The following table shows examples of algebraic symbolism for addition. Just as in arithmetic expressions, use the plus (+) symbol for addition. The letters x and y are used in the table to represent unknown numbers.

Algebraic Symbolism for Addition	
Sample Word Phrase	**Symbolism**
The sum of a number and 6	$x + 6$
The sum of 6 and a number	$6 + x$
8 plus a number	$8 + x$
A number plus 8	$x + 8$
A number added to 4	$4 + x$
4 added to a number	$x + 4$
A number increased by 10	$x + 10$
10 increased by a number	$10 + x$
5 more than a number	$x + 5$
A number exceeded by 2	$x + 2$
The sum of x and y	$x + y$

The table illustrates that addition expressions such as $x + 6$ can be read in any one of several ways. Without the context of a problem, read $x + 6$ as "the sum of x and 6" or "x plus 6."

Use variables to write expressions for problems involving addition. Here are examples.

> Bethany's age at the present time is unknown. Write an expression for Bethany's age 5 years from now.

Let b = Bethany's present age in years. Then, $b + 5$ years = Bethany's age 5 years from now.

> The original price of an item is unknown. Write an expression to show an increase of \$10.45 in the original price.

Let c = the original price of the item in dollars. Then, $c + \$10.45$ = the original price increased by \$10.45.

☞ Try These

1. Fill in the blank. The variables x and y represent specific numbers.
   - **(a)** The symbol for addition in algebraic expressions is __________.
   - **(b)** In the expression $x + 5$, x represents a __________.
   - **(c)** The expression $6 + y$ represents a __________.
   - **(d)** The expression $x + y$ represents a __________.

2. Write an expression for the given phrase.
   - **(a)** the sum of a number and 12
   - **(b)** a number exceeded by 4.5
   - **(c)** 200 more than a number
   - **(d)** $\frac{1}{4}$ plus a number
   - **(e)** y added to x

3. Write the given expression in words.

   (a) $x + 4$
   (b) $y + 10$

4. Write an expression for the given problem.

   (a) Diwan's age at the present time is unknown. Write an expression for Diwan's age 10 years from now.
   (b) The original price of an item is unknown. Write an expression to show an increase of $16.92 above the original price.
   (c) The width of a rectangle is unknown. The length of the rectangle is 4 feet more than its width. Write an expression for the length of the rectangle.
   (d) The number of girls in the classroom is unknown. The number of boys is 6 more than the number of girls. Write an expression for the number of boys in the room.
   (e) The distance to City A from Mira's house is unknown. The distance to City B from Mira's house exceeds the distance to City A by 50 miles. Write an expression for the distance to City B from Mira's house.

Solutions

1. **(a)** +
   (b) number
   (c) number
   (d) number
2. **(a)** $x + 12$
   (b) $x + 4.5$
   (c) $x + 200$
   (d) $\frac{1}{4} + x$
   (e) $x + y$
3. **(a)** "the sum of x and 4" or "x plus 4"
   (b) "the sum of y and 10" or "$y + 10$"
4. **(a)** Let d = Diwan's present age in years. Then, d + 10 years = Diwan's age 10 years from now.
   (b) Let c = the original price of the item in dollars. Then, c + \$16.92 = the original price increased by \$16.92.
   (c) Let w = the width of the rectangle in feet. Then, w + 4 feet = the length of the rectangle.
   (d) Let g = the number of girls in the classroom. Then, $g + 6$ = the number of boys in the classroom.
   (e) Let a = the distance to City A from Mira's house in miles. Then, a + 50 miles = the distance to City B from Mira's house.

Writing and Reading Subtraction Expressions

The following table shows examples of algebraic symbolism for subtraction. Just as in arithmetic expressions, use the minus (–) symbol for subtraction. The letters x and y are used in the table to represent unknown numbers.

Algebraic Symbolism for Subtraction	
Sample Word Phrase	**Symbolism**
The difference of a number and 6	$x - 6$
The difference of 6 and a number	$6 - x$
8 minus a number	$8 - x$
A number minus 8	$x - 8$
A number subtracted from 4	$4 - x$
4 subtracted from a number	$x - 4$
A number decreased by 10	$x - 10$
10 decreased by a number	$10 - x$
5 less than a number	$x - 5$
A number reduced by 2	$x - 2$
The difference of x and y	$x - y$
The difference of y and x	$y - x$

Tip: The order of the numbers in a subtraction problem is important. In general, (x minus y) $\neq$ (y minus x). (The symbol $\neq$ means "is not equal to.")

The table illustrates that subtraction expressions such as $x - 6$ can be read in any one of several ways. Without the context of a problem, read $x - 6$ as "the difference of x and 6" or as "x minus 6."

Use variables to write expressions for problems involving subtraction. Here are examples.

Johann's age at the present time is unknown. Write an expression for Johann's age 8 years ago.

Let j = Johann's present age in years. Then, $j - 8$ years = Johann's age 8 years ago.

The selling price of an item is unknown. Write an expression to show a discount of \$15 off the selling price.

Tip: A discount is an amount taken off the price of an item. It represents a decrease in the price.

Let s = the selling price of the item in dollars. Then, s – \$15 = the selling price decreased by \$15.

☞Try These

1. Fill in the blank. The variables x and y represent specific numbers.
   - **(a)** The symbol for subtraction in algebraic expressions is __________.
   - **(b)** The expression $x - 6$ represents a __________.
   - **(c)** The expression $10 - x$ represents a __________.
   - **(d)** The expression $x - y$ represents a __________.

2. Write an expression for the given phrase.

   (a) the difference of a number and 9
   (b) 7.25 decreased by a number
   (c) 34 less than a number
   (d) a number minus $\frac{2}{3}$
   (e) y subtracted from x

3. Write the given expression in words.

   (a) $x - 4$
   (b) $y - 10$

4. Write an expression for the given problem.

   (a) Rita's age at the present time is unknown. Write an expression for Rita's age 4 years ago.
   (b) The selling price of an item is unknown. Write an expression to show a discount of $14 off the selling price.
   (c) The length of a rectangle is unknown. The width of the rectangle is 20 centimeters less than its length. Write an expression for the width of the rectangle.
   (d) The number of black marbles in the box is unknown. The number of red marbles is 12 less than the number of black marbles. Write an expression for the number of red marbles in the box.
   (e) The distance Paul ran is unknown. Antonio ran 5 meters less than Paul ran. Write an expression for the distance Antonio ran.

Solutions

1. **(a)** –
   (b) number
   (c) number
   (d) number
2. **(a)** $x - 9$
   (b) $7.25 - x$
   (c) $x - 34$
   (d) $x - \frac{2}{3}$
   (e) $x - y$
3. **(a)** "the difference of x and 4" or "x minus 4"
   (b) "the difference of y and 10" or "y minus 10"
4. **(a)** Let r = Rita's present age in years. Then, $r - 4$ years = Rita's age 4 years ago.
   (b) Let s = the selling price of the item in dollars. Then, $s - \$14$ = the selling price decreased by $14.
   (c) Let l = the length of the rectangle in centimeters. Then, $l - 20$ centimeters = the width of the rectangle.
   (d) Let b = the number of black marbles in the box. Then, $b - 12$ = the number of red marbles in the box.
   (e) Let d = the distance Paul ran in meters. Then, $d - 5$ meters = the distance Antonio ran.

Writing and Reading Multiplication Expressions

When you use letters in expressions, do not use the times (×) symbol for multiplication. The × symbol can be confused with the letter x. To show 5 times x, do one of the following:

- Use the dot (·) multiplication symbol between the factors, like this: $5 \cdot x$. Read $5 \cdot x$ as "five times x" or as "the product of 5 and x."
- Enclose one or both factors in parentheses, like this: $(5)(x)$, $5(x)$, or $(5)x$. Read each of these as "five times x" or as "the product of 5 and x."
- Write the factors side by side with no symbols between them, like this: $5x$. Read $5x$ as "five x." The expression $5x$ is the result of multiplying 5 times x. So $5 \cdot x = 5x$, $(5)(x) = 5x$, $5(x) = 5x$, and $(5)x = 5x$.

Tip: If no letters are involved in a product, as in 7 times 4, write 7 · 4 or (7)(4) or 7 × 4, but not 74. Writing the factors 7 and 4 side by side looks like the number 74 instead of the product, 7 times 4.

In the product of a constant and a variable, the constant factor is the variable's **numerical coefficient.** For example, 5 is the numerical coefficient of x in the expression $5x$. If no number is shown as a factor times a variable, then the numerical coefficient of the variable is 1. For example, the numerical coefficient of y is 1.

Tip: When you use the side-by-side form for the product of a number and a variable, write the number first, as in 5*x*.

The following table shows examples of algebraic symbolism for multiplication. The letters x and y are used in the table to represent unknown numbers.

Algebraic Symbolism for Multiplication	
Sample Word Phrase	**Symbolism**
The product of 6 and a number	$6x$, $6 \cdot x$, $(6)(x)$, $6(x)$, or $(6)x$
8 times a number	$8x$, $8 \cdot x$, $(8)(x)$, $8(x)$, or $(8)x$
A number multiplied by 4	$4x$, $4 \cdot x$, $(4)(x)$, $4(x)$, or $(4)x$
10 of a number	$10x$, $10 \cdot x$, $(10)(x)$, $10(x)$, or $(10)x$
$\frac{3}{4}$ of a number	$\frac{3}{4}x$, $\frac{3}{4} \cdot x$, $\left(\frac{3}{4}\right)(x)$, $\frac{3}{4}(x)$, or $\left(\frac{3}{4}\right)x$
1.5 of a number	$1.5x$, $1.5 \cdot x$, $(1.5)(x)$, $1.5(x)$, or $(1.5)x$
5% of a number	$5\%x$, $5\% \cdot x$, $(5\%)(x)$, $5\%(x)$, or $(5\%)x$
Half a number	$\frac{1}{2}x$, $\frac{1}{2} \cdot x$, $\left(\frac{1}{2}\right)(x)$, $\frac{1}{2}(x)$, or $\left(\frac{1}{2}\right)x$
Twice a number	$2x$, $2 \cdot x$, $(2)(x)$, $2(x)$, or $(2)x$
Double a number	$2x$, $2 \cdot x$, $(2)(x)$, $2(x)$, or $(2)x$
Triple a number	$3x$, $3 \cdot x$, $(3)(x)$, $3(x)$, or $(3)x$
Quadruple a number	$4x$, $4 \cdot x$, $(4)(x)$, $4(x)$, or $(4)x$
The product of x and y	xy, $x \cdot y$, $(x)(y)$, $x(y)$, or $(x)y$
The product of y and x	yx, $y \cdot x$, $(y)(x)$, $y(x)$, or $(y)x$
Three times the product of x and y	$3(xy)$ or $3(x \cdot y)$

Tip: Use parentheses to make your meaning clear.

Use letters to write expressions for problems involving multiplication. Here are examples.

Taylor's age is unknown. Vi's age is triple Taylor's age. Write an expression for Vi's age.

Let t = Taylor's age in years. Then, $3t$ = Vi's age in years.

The selling price of an item is unknown. The item is marked 20% off. When an item is marked 20% off, the customer saves 20% of the selling price. Write an expression to show 20% of the selling price.

Let s = the selling price of the item in dollars. Then $20\% s$ = 20% of the selling price.

Try These

1. Fill in the blank.

   (a) The expression $5x$ is the result of __________ 5 and x.
   (b) The numerical coefficient of w is __________.
   (c) The numerical coefficient of $25b$ is __________.
   (d) The numerical coefficient of $\frac{1}{4}y$ is __________.
   (e) The numerical coefficient of $1.25a$ is __________.

2. Write an expression for the given phrase five different ways.

   (a) the product of 12 and a number
   (b) 4.5 of a number
   (c) 60 times a number
   (d) a number multiplied by $\frac{3}{4}$
   (e) triple a number

3. Write the given expression in words.

   (a) $7x$
   (b) $y \cdot 9$
   (c) $(z)(20)$

4. Write an expression for the given problem.

   (a) Ismael's age is unknown. His father's age is double Ismael's age. Write an expression for Ismael's father's age.
   (b) The selling price of an item is unknown. The item is marked 10% off. Write an expression to show 10% of the selling price.
   (c) The formula for the area, A, of a rectangle is $A = lw$, where l is the rectangle's length and w is its width. The width of a rectangle is unknown. The length of the rectangle is 10 feet. Write an expression for the area of the rectangle.
   (d) The number of girls in the club is unknown. The number of boys is $\frac{3}{4}$ the number of girls. Write an expression for the number of boys in the club.
   (e) The distance to City A from Indio's house is unknown. The distance to City B from Indio's house is 1.5 times the distance to City A. Write an expression for the distance to City B from Indio's house.

Solutions

1. **(a)** multiplying
   (b) 1
   (c) 25
   (d) $\frac{1}{4}$
   (e) 1.25
2. **(a)** $12x$, $12 \cdot x$, $(12)(x)$, $12(x)$, or $(12)x$
   (b) $4.5x$, $4.5 \cdot x$, $(4.5)(x)$, $4.5(x)$, or $(4.5)x$
   (c) $60x$, $60 \cdot x$, $(60)(x)$, $60(x)$, or $(60)x$
   (d) $\frac{3}{4}x$, $\frac{3}{4} \cdot x$, $\left(\frac{3}{4}\right)(x)$, $\frac{3}{4}(x)$, or $\left(\frac{3}{4}\right)x$
   (e) $3x$, $3 \cdot x$, $(3)(x)$, $3(x)$, or $(3)x$
3. **(a)** "seven x," "7 times x," or "the product of 7 and x"
   (b) "y times 9" or "the product of y and 9"
   (c) "z times 20" or "the product of z and 20"
4. **(a)** Let x = Ismael's age in years. Then, $2x$ = Ismael's father's age in years.
   (b) Let s = the selling price of the item in dollars. Then $10\% s$ = 10% of the selling price.
   (c) Let w = the width of the rectangle. Then, $A = lw = (10 \text{ feet})(w)$.
   (d) Let g = the number of girls in the club. Then, $\frac{3}{4} \cdot g$ = the number of boys in the club.
   (e) Let a = the distance to City A from Indio's house. Then, $1.5a$ = the distance to City B from Indio's house.

Writing and Reading Division Expressions

When you use letters in expressions, do not use the symbols ÷ or $\overline{)}$ for division. Instead, for example, write a divided by b as $\frac{a}{b}$. Without the context of a problem, read $\frac{a}{b}$ as "a divided by b" or as "the quotient of a and b."

Tip: dividend ÷ divisor = $\frac{\text{dividend}}{\text{divisor}}$ and divisor$\overline{)\text{dividend}}$ = $\frac{\text{dividend}}{\text{divisor}}$.

Remember, in division the *divisor* CANNOT be zero. Division by zero has no meaning. You say $\frac{a}{0}$ is "undefined." For example, $\frac{10}{0}$ is undefined. But if the *dividend* is zero, the quotient is 0. The quotient $\frac{0}{b}$ has meaning and is defined. Zero divided by any nonzero number is 0. For example, $\frac{0}{10}$ is 0.

Tip: NEVER put 0 in the denominator of a fraction.

The following table shows examples of algebraic symbolism for division. The letters x and y are used in the table to represent unknown nonzero numbers.

Algebraic Symbolism for Division	
Sample Word Phrase	**Symbolism**
The quotient of a number and 6	$\frac{x}{6}$
The quotient of 6 and a number	$\frac{6}{x}$
A number divided by 8	$\frac{x}{8}$
8 divided by a number	$\frac{8}{x}$
The ratio of a number to 4	$\frac{x}{4}$
The ratio of 4 to a number	$\frac{4}{x}$
x for every 5	$\frac{x}{5}$
x per 1	$\frac{x}{1}$
x for each 10	$\frac{x}{10}$
The quotient of *x* and *y*	$\frac{x}{y}$
The quotient of *y* and *x*	$\frac{y}{x}$
The quotient of 0 and *x*	$\frac{0}{x}=0$
The quotient of *x* and 0	Undefined, because $\frac{x}{0}$ has no meaning

Tip: The order of the numbers in a division problem is important. In general, (*x* divided by *y*) ≠ (*y* divided by *x*).

Use letters to write expressions for problems involving division. Here are examples.

The cost of the pizza is unknown. Six girls share equally the cost of the pizza. Write an expression for the amount in dollars each girl pays.

Let c = the cost of the pizza in dollars. Then, $\frac{c}{6}$ = amount in dollars each girl pays.

The distance traveled in miles on the trip is unknown. The trip takes 3 hours. Write an expression for the ratio of the distance traveled to the time for the trip.

Let d = the distance traveled in miles. Then, $\frac{d}{3\text{ hours}}$ = the ratio of the distance traveled to the time for the trip.

☞ Try These

1. Write the division expression as a fraction.

   (a) $24 \div 3$
   (b) $5\overline{)75}$
   (c) $4.5 \div 1.5$
   (d) $5\overline{)4}$
   (e) $4\overline{)5}$

2. Write an expression for the given phrase.

   (a) the quotient of a number and 12
   (b) a number divided by 0.5
   (c) the ratio of a number to 100
   (d) the quotient of a number and 3.2
   (e) 15 divided by a number

3. Write the given expression in words.

   (a) $\frac{x}{30}$
   (b) $\frac{y}{3}$

4. Write an expression for the given problem.

   (a) The cost of the pizza is unknown. Three boys share equally the cost of the pizza. Write an expression for the amount in dollars each boy pays.
   (b) The distance traveled in miles on a trip is unknown. The trip takes 2.5 hours. Write an expression for the ratio of the distance traveled to the time for the trip.
   (c) The number of students is unknown. The students are put into three equal-sized groups. Write an expression for the number of students in each group.
   (d) The number of beads is unknown. The beads are divided equally among 15 children. Write an expression for the number of beads each child gets.
   (e) The price of four cans of soup is unknown. Write an expression that can be used to find the price of one can of soup.

Solutions

1. **(a)** $\frac{24}{3}$
   (b) $\frac{75}{5}$
   (c) $\frac{4.5}{1.5}$
   (d) $\frac{4}{5}$
   (e) $\frac{5}{4}$

2. **(a)** $\frac{x}{12}$
   (b) $\frac{x}{0.5}$
   (c) $\frac{x}{100}$
   (d) $\frac{x}{3.2}$
   (e) $\frac{15}{x}$
3. **(a)** "x divided by 30" or "the quotient of x and 30"
   (b) "y divided by 3" or "the quotient of y and 3"
4. **(a)** Let c = the cost of the pizza in dollars. Then, $\frac{c}{3}$ = the amount in dollars each boy pays.
   (b) Let d = the distance traveled in miles. Then, $\frac{d}{2.5 \text{ hours}}$ = ratio of the distance traveled to the time for the trip.
   (c) Let s = the number of students. Then, $\frac{s}{3}$ = the number of students in each group.
   (d) Let b = the number of beads. Then, $\frac{b}{15}$ = the number of beads each child gets.
   (e) Let p = the price of four cans of soup. Then, $\frac{p}{4}$ can be used to find the price of one can of soup.

Evaluating Expressions at Specific Values of Their Variables

(CCSS.Math.Content.6.EE.A.2.C)

You don't know what number an algebraic expression represents unless you know what number (or numbers) the variable (or variables) in the expression stand for. However, if you are given numerical values for the variables, you can evaluate the algebraic expression by substituting the given numerical value for each variable and then performing the indicated operations. Just be sure to *follow the order of operations* as you proceed. Here are examples.

Find the value of $9x$ when $x = 4$.

Put parentheses around x: $9(x)$.

Replace x with 4, and then evaluate: $9(4) = 36$.

Tip: When x is 4, $9x$ is *not* 94. The term $9x$ means "9 times x." When x is 4, $9x$ is 9 times 4, or 36.

Find the value of $24 + 2x^2$ when $x = 7$.

Put parentheses around x: $24 + 2(x)^2$.

Replace x with 7, and then evaluate: $24 + 2(7)^2 = 24 + 2(49) = 24 + 98 = 122$.

Find the value of $2.5a + \frac{3a}{15} - 5$ when $a = 10$.

Put parentheses around a: $2.5(a)+\frac{3(a)}{15}-5$.

Replace a with 10, and then evaluate: $2.5(10)+\frac{3(10)}{15}-5=25+\frac{30}{15}-5=25+2-5=22$.

Find the value of $5x-2y$ when $x=8$ and $y=12$.

Put parentheses around x and y: $5(x)-2(y)$.

Replace x with 8 and y with 12, and then evaluate: $5(8)-2(12)=40-24=16$.

Find the value of $\frac{1}{2}(20+3b)$ when $b=14$.

Put parentheses around b: $\frac{1}{2}(20+3(b))$.

Replace b with 14, and then evaluate: $\frac{1}{2}(20+3(14))=\frac{1}{2}(20+42)=\frac{1}{2}(62)=31$.

A **formula** expresses a rule about a relationship between quantities represented by symbols. Formulas are used in numerous areas in the real world. Use your skills in evaluating algebraic expressions to evaluate formulas. If the replacements for the variables in formulas have units, include the units in the computations when it makes sense to do so. Here is an example.

The formula for the perimeter, P, of a rectangle is $P=2l+2w$, where l is the length of the rectangle and w is its width. Find the perimeter of a rectangle that has length $l=12$ feet and width $w=9$ feet.

Put parentheses around l and w in the formula: $P=2(l)+2(w)$.

Replace l with 12 feet and w with 9 feet, and then evaluate: $P=2(12\text{ feet})+2(9\text{ feet})=24\text{ feet}+18\text{ feet}=$ 42 feet.

Tip: After you feel comfortable replacing variables with their numerical values, you can skip the step of putting parentheses around the variables first.

☞ Try These

1. Find the value of the algebraic expression when $x=12$ and $y=5$.

   (a) $4x+25$
   (b) $2xy$
   (c) $3x-2y$
   (d) $y(x+8)$
   (e) $\frac{7x}{6}$

2. Find the value of the algebraic expression when $a = 4.5$ and $b = 10$.

   (a) $2a - 5$
   (b) $4(b - a)$
   (c) $\frac{1}{2}b^2$
   (d) $2ab$
   (e) $3a + 2.5b$

3. The formula for the perimeter, P, of a rectangle is $P = 2l + 2w$, where l is the length of the rectangle and w is its width. Find the perimeter of a rectangle that has length $l = 15$ units and width $w = 8$ units.

4. The formula for converting Fahrenheit temperature to Celsius temperature is $C = \frac{5}{9}(F - 32°)$, where F is Fahrenheit temperature in degrees. Find the Celsius temperature that corresponds to 86°F.

Solutions

1. **(a)** $4x + 25 = 4(x) + 25 = 4(12) + 25 = 48 + 25 = 73$
   (b) $2xy = 2(x)(y) = 2(12)(5) = 120$
   (c) $3x - 2y = 3(x) - 2(y) = 3(12) - 2(5) = 36 - 10 = 26$
   (d) $y(x + 8) = 5(12 + 8) = 5(20) = 100$
   (e) $\frac{7x}{6} = \frac{7(x)}{6} = \frac{7(12)}{6} = \frac{84}{6} = 14$
2. **(a)** $2a - 5 = 2(4.5) - 5 = 9 - 5 = 4$
   (b) $4(b - a) = 4(10 - 4.5) = 4(5.5) = 22.0 = 22$
   (c) $\frac{1}{2}b^2 = \frac{1}{2}(10)^2 = \frac{1}{2}(100) = 50$
   (d) $2ab = 2(4.5)(10) = 90$
   (e) $3a + 2.5b = 3(4.5) + 2.5(10) = 13.5 + 25 = 38.5$
3. $P = 2l + 2w = 2(15 \text{ units}) + 2(8 \text{ units}) = 30 \text{ units} + 16 \text{ units} = 46 \text{ units}$
4. $C = \frac{5}{9}(F - 32°) = \frac{5}{9}(86° - 32°) = \frac{5}{9}(54°) = \frac{5}{\cancel{9}_1}\left(\frac{\cancel{54}^{6}\,°}{1}\right) = 30°\text{C}$

Identifying and Generating Equivalent Expressions

(CCSS.Math.Content.6.EE.A.3, CCSS.Math.Content.6.EE.A.4)

Equivalent expressions evaluate to the same number, no matter what number replaces the variable in each expression. Or if there is more than one variable, there is a replacement number for each variable. Here are examples.

The expressions $a + 5$ and $5 + a$ are equivalent expressions. Every replacement for a will yield the same number for both $a + 5$ and $5 + a$. If $a = 10$, then $a + 5 = 10 + 5 = 15$ and $5 + a = 5 + 10 = 15$, the same number. If $a = 7.5$, then $a + 5 = 7.5 + 5 = 12.5$ and $5 + a = 5 + 7.5 = 12.5$, the same number. If $a = 6\frac{3}{4}$, then $a + 5 = 6\frac{3}{4} + 5 = 11\frac{3}{4}$ and $5 + a = 5 + 6\frac{3}{4} = 11\frac{3}{4}$, the same number, and so forth.

The expressions xy and yx are equivalent expressions. No matter what replacements you use for x and y, xy and yx will evaluate to the same number. If $x = 4$ and $y = 7$, then $xy = 4 \cdot 7 = 28$ and $yx = 7 \cdot 4 = 28$, the same number. If $x = 8.3$ and $y = 10$, then $xy = (8.3)(10) = 83$ and $yx = (10)(8.3) = 83$, the same number. If $x = \frac{1}{3}$ and $y = 18$, then $xy = \left(\frac{1}{3}\right)(18) = 6$ and $yx = (18)\left(\frac{1}{3}\right) = 6$, the same number, and so forth.

You can identify equivalent expressions by recognizing them as the result of applying one or more properties of the operations of addition and multiplication for numbers. For instance, the commutative property of addition guarantees $a + 5$ and $5 + a$ are always equal no matter what number replaces a. Similarly, the commutative property of multiplication guarantees xy and yx are always equal no matter what numbers replace x and y.

The following table summarizes the commutative and associative properties for addition and multiplication.

Commutative and Associative Properties

Operation	Property	Rule	Example
Addition	Commutative (any order)	$a + b = b + a$	$4 + 9 = 13$ and $9 + 4 = 13$
Multiplication	Commutative (any order)	$a \cdot b = b \cdot a$	$7 \cdot 6 = 42$ and $6 \cdot 7 = 42$
Addition	Associative (any grouping)	$(a + b) + c = a + (b + c)$	$(3 + 1) + 5 = 4 + 5 = 9$ and $3 + (1 + 5) = 3 + 6 = 9$
Multiplication	Associative (any grouping)	$(a \cdot b) \cdot c = a \cdot (b \cdot c)$	$(2 \cdot 3) \cdot 7 = 6 \cdot 7 = 42$ and $2 \cdot (3 \cdot 7) = 2 \cdot 21 = 42$

The distributive property involves both addition and multiplication. In words, "multiplication distributes over addition." The following table summarizes useful forms of the distributive property.

Forms of the Distributive Property

Form Description	Rule	Example
Product to sum	$a(b + c) = a \cdot b + a \cdot c = ab + ac$	$3(20 + 7) = 3(27) = 81$ and $3 \cdot 20 + 3 \cdot 7 = 60 + 21 = 81$
Product to sum	$(b + c)a = b \cdot a + c \cdot a = ba + ca$	$(10 + 3)5 = (13)5 = 65$ and $10 \cdot 5 + 3 \cdot 5 = 50 + 15 = 65$
Sum to product	$ab + ac = a \cdot b + a \cdot c = a(b + c)$	$4 \cdot 2 + 4 \cdot 3 = 8 + 12 = 20$ and $4(2 + 3) = 4(5) = 20$
Sum to product	$ba + ca = b \cdot a + c \cdot a = (b + c)a$	$3 \cdot 5 + 7 \cdot 5 = 15 + 35 = 50$ and $(3 + 7)5 = (10)5 = 50$

The distributive property can be extended to include the sum of more than two numbers. Here is an example.

$$4(32 + 53) = 4(30 + 2 + 50 + 3) = 4 \cdot 30 + 4 \cdot 2 + 4 \cdot 50 + 4 \cdot 3 = 120 + 8 + 200 + 12 = 340$$

The properties for the operations hold true when one or more variables hold the place for numbers. Here are examples.

$18a + 12b = 6 \cdot 3a + 6 \cdot 2b = 6(3a + 2b)$

Tip: Notice 6 is the GCF for $18a$ and $12b$.

$5x + 3x = (5 + 3)x = 8x$

Tip: Notice x is the GCF for $5x$ and $3x$.

Generate equivalent variable expressions by using properties of the operations of addition and multiplication. Here are examples.

$20 + n$ is equivalent to $n + 20$ (by the commutative property of addition)

$\left(\frac{1}{2}\right)(14y) = \left(\frac{1}{2} \cdot 14\right)(y) = (7)(y) = 7y$ (by the associative property of multiplication)

$3(1 + 2x) = 3 \cdot 1 + 3 \cdot 2x = 3 + 6x$ (by the distributive property)

$x + x + x = 1x + 1x + 1x = (1 + 1 + 1)x = (3)x = 3x$ (by the distributive property)

Tip: Remember, if no number is shown as a factor times a variable, then the variable's numerical coefficient is 1. So the numerical coefficient of x is 1.

To show two variable expressions are *not* equivalent, substitute a number that makes the two expressions evaluate to different numbers. ***Tip:*** The numbers 0 and 1 are usually not good values to use when testing whether expressions are equivalent. If the expressions have more than one variable, substitute a number for each variable. Here are examples.

Is $20 + n$ equivalent to $20n$?

No, because if $n = 2$, $20 + n = 20 + 2 = 22$, but $20n = 20(2) = 40$.

Are $3(7x)$ and $10x$ equivalent expressions?

No, because if $x = 2$, $3(7x) = 3(7 \cdot 2) = 3(14) = 42$, but $10x = 10 \cdot 2 = 20$.

Are $2x + y$ and $2xy$ equivalent expressions?

No, because if $x = 10$ and $y = 5$, $2x + y = 2(10) + 5 = 25$, but $2xy = 2(10)(5) = 100$.

☞ Try These

1. Are the given expressions equivalent? Justify your answer.

   (a) $8(2x)$ and $10x$
   (b) $24x + 18y$ and $6(4x + 3y)$
   (c) $a + b + a + b$ and $2a + 2b$
   (d) $5x + 6x$ and $30x$
   (e) $4 + 3n$ and $7n$

2. Use operation properties to write a sum equivalent to the given product.

   (a) $5(x + y)$
   (b) $2(3x + 3.5)$
   (c) $\frac{1}{2}(10w + 16)$
   (d) $100(0.14m + 1.25)$
   (e) $15\left(\frac{2}{3}y + \frac{4}{5}\right)$

3. Use operation properties to write a product equivalent to the given sum.

   (a) $8 + 56a$
   (b) $6y + 21z$
   (c) $9x + 4x$
   (d) $m + m + m + m + m$
   (e) $30x + 18$

4. Verify the expressions are equivalent when $x = \frac{1}{2}$ and $y = 2.4$.

   (a) $8x + 6x$ and $14x$
   (b) $x + y + x$ and $2x + y$
   (c) $5y + 25$ and $25 + 5y$
   (d) $24x + 15y$ and $3(8x + 5y)$
   (e) $5(2x + 3)$ and $10x + 15$

Solutions

1. **(a)** No, because if $x = 2$, $8(2x) = 8(2 \cdot 2) = 8(4) = 32$, but $10x = 10 \cdot 2 = 20$.
   (b) Yes, $24x + 18y = 6 \cdot 4x + 6 \cdot 3y = 6(4x + 3y)$ (by the distributive property).
   (c) Yes, $a + b + a + b = a + a + b + b$ (by the associative property of addition) $= 1a + 1a + 1b + 1b$ (by the meaning of coefficient) $= (1 + 1)a + (1 + 1)b$ (by the distributive property) $= (2)a + (2)b = 2a + 2b$.
   (d) No, because if $x = 2$, $5x + 6x = 5(2) + 6(2) = 10 + 12 = 22$, but $30x = 30 \cdot 2 = 60$.
   (e) No, because if $n = 2$, $4 + 3n = 4 + 3(2) = 4 + 6 = 10$, but $7n = 7 \cdot 2 = 14$. ***Tip:*** If $n = 1$, $4 + 3n = 4 + 3(1) = 4 + 3 = 7$ and $7n = 7 \cdot 1 = 7$, the same number. This result does not mean $4 + 3n$ and $7n$ are equivalent. To be equivalent expressions, they must evaluate to the same number for all replacements for n. They do not evaluate to the same number when $n = 2$. ***Tip:*** This problem illustrates that 0 and 1 are usually not good values to use when testing whether expressions are equivalent.
2. **(a)** $5(x + y) = 5 \cdot x + 5 \cdot y = 5x + 5y$
   (b) $2(3x + 3.5) = 2 \cdot 3x + 2 \cdot 3.5 = 6x + 7$
   (c) $\frac{1}{2}(10w + 16) = \frac{1}{2} \cdot 10w + \frac{1}{2} \cdot 16 = \left(\frac{1}{2} \cdot 10\right)w + \left(\frac{1}{2} \cdot 16\right) = (5)w + (8) = 5w + 8$
   (d) $100(0.14m + 1.25) = 100 \cdot 0.14m + 100 \cdot 1.25 = (100 \cdot 0.14)m + (100 \cdot 1.25) = 14m + 125$
   (e) $15\left(\frac{2}{3}y + \frac{4}{5}\right) = 15 \cdot \frac{2}{3}y + 15 \cdot \frac{4}{5} = \left(15 \cdot \frac{2}{3}\right)y + \left(15 \cdot \frac{4}{5}\right) = (10)y + (12) = 10y + 12$

3. **(a)** $8 + 56a = 8 \cdot 1 + 8 \cdot 7a = 8(1 + 7a)$
(b) $6y + 21z = 3 \cdot 2y + 3 \cdot 7z = 3(2y + 7z)$
(c) $9x + 4x = (9 + 4)x = 13x$
(d) $m + m + m + m + m = 1m + 1m + 1m + 1m + 1m = (1 + 1 + 1 + 1 + 1)m = 5m$
(e) $30x + 18 = 6 \cdot 5x + 6 \cdot 3 = 6(5x + 3)$

4. **(a)** $8x + 6x = 8 \cdot \frac{1}{2} + 6 \cdot \frac{1}{2} = 4 + 3 = 7$ and $14x = 14 \cdot \frac{1}{2} = 7$
(b) $x + y + x = \frac{1}{2} + 2.4 + \frac{1}{2} = 3.4$ and $2x + y = 2 \cdot \frac{1}{2} + 2.4 = 1 + 2.4 = 3.4$
(c) $5y + 25 = 5(2.4) + 25 = 12 + 25 = 37$ and $25 + 5y = 25 + 5(2.4) = 25 + 12 = 37$
(d) $24x + 15y = 24 \cdot \frac{1}{2} + 15(2.4) = 12 + 36 = 48$ and $3(8x + 5y) = 3\left(8 \cdot \frac{1}{2} + 5 \cdot 2.4\right) = 3(4 + 12) = 3(16) = 48$
(e) $5(2x + 3) = 5\left(2 \cdot \frac{1}{2} + 3\right) = 5(1 + 3) = 5(4) = 20$ and $10x + 15 = 10 \cdot \frac{1}{2} + 15 = 5 + 15 = 20$

Solving One-Step Single-Variable Mathematical and Real-World Equations

(CCSS.Math.Content.6.EE.B.5, CCSS.Math.Content.6.EE.B.6, CCSS.Math.Content.6.EE.B.7)

An **equation** is a statement that two mathematical expressions are equal. Here are examples.

$$x + 10.25 = 106.24,\ x - \frac{3}{4} = \frac{1}{2},\ 9x = 11.25,\ \frac{x}{4} = 14$$

An equation has two sides. Whatever is on the left side of the equal sign is the *left side* of the equation, and whatever is on the right side of the equal sign is the *right side* of the equation.

Equations that contain only numbers are either true or false. The equation $8 + 5 = 13$ is true. But the equation $0 = 2$ is false.

Variables may hold places for numbers in an equation. The equation $x + 8 = 18$ has one variable, namely x. When a variable holds the place for a number in an equation, the equation is **open.** The equation $x + 8 = 18$ is an open equation that has one variable, x. Without knowing what number x is standing for, you cannot say whether $x + 8 = 18$ is true or false. Replace x with 5. Then $x + 8 = 18$ is false because $5 + 8 = 13$, not 18. Replace x with 10. Then $x + 8 = 18$ is true because $10 + 8 = 18$. In fact, 10 is the only replacement for x that makes $x + 8 = 18$ a true statement.

To **solve an equation** that has one variable means to find a number replacement for the variable that makes the equation a true statement. An equation is true when the left side has the same value as the right side. A **solution** to an equation is a number that, when substituted for the variable, makes the equation true. For example, 10 is a solution to the equation $x + 8 = 18$. It is a number that, when substituted for x, makes the equation $x + 8 = 18$ a true statement. Equations that have the same solution are **equivalent.**

To determine whether a number is a solution to an equation, replace the variable with the number and perform the operations indicated on each side of the equation. If this results in a true equation, the number is a solution. This process is called **checking** a solution. For example, 10 is a solution to the equation $x + 8 = 18$ because $10 + 8 = 18$ is a true statement.

In this section, you will solve one-step, one-variable equations. A one-step, one-variable equation has the form $x + p = q$, $x - p = q$, $px = q$, or $\frac{x}{p} = q$. To **solve** a one-step, one-variable equation, undo the operation that has been done to the variable to obtain an equivalent equation that has the form: variable = solution. The most important rule to remember when you are solving an equation is that *whatever you do to one side of the equation you must do to the other side of the equation.* Think of the equal sign as a balance point. To keep the equation in balance, you do the same thing to both sides of the equation.

The main actions that will result in equivalent equations are

- Adding the same number to both sides of the equation
- Subtracting the same number from both sides of the equation
- Multiplying both sides of the equation by the same nonzero number
- Dividing both sides of the equation by the same nonzero number

What has been done to the variable determines the operation you choose to do. You do it to both sides of the equation to keep the equation balanced. If something has been added to the variable, as in $x + p = q$, subtraction will undo the addition. If something has been subtracted from the variable, as in $x - p = q$, addition will undo the subtraction. If the variable has been multiplied by a number, as in $px = q$, division will undo the multiplication. And if the variable has been divided by a number, as in $\frac{x}{p} = q$, multiplication will undo the division.

☞ Try These

1. Fill in the blank.

   (a) An equation is a statement that two mathematical expressions are __________.

   (b) An equation has two __________.

   (c) When a variable holds the place for a number in an equation, the equation is __________.

   (d) To solve an equation that has one variable means to find a number replacement for the variable that makes the equation __________.

   (e) A __________ to an equation is a number that, when substituted for the variable, makes the equation true.

2. What should you do to the given equation to solve it?

   (a) $x - 4.5 = 12.6$

   (b) $5x = 6.45$

   (c) $x + \frac{3}{2} = \frac{11}{4}$

   (d) $\frac{x}{2} = 5.75$

Solutions

1. **(a)** equal
   (b) sides
   (c) open
   (d) true
   (e) solution
2. **(a)** Add 4.5 to both sides of the equation.
   (b) Divide both sides of the equation by 5.
   (c) Subtract $\frac{3}{2}$ from both sides of the equation.
   (d) Multiply both sides of the equation by 2.

Solving One-Step Mathematical Equations of the Forms $x + p = q$ and $x - p = q$

Suppose you have the following problem.

When 10.25 is added to a number, the sum is 106.24. What is the number?

In arithmetic, to find the value of the number, you should decide to subtract 10.25 from 106.24. When you do the subtraction, you get $106.24 - 10.25 = 95.99$.

Here is the way to write the problem using an equation.

Solve $x + 10.25 = 106.24$.

Think: *"10.25 is added to x, a number whose value I do not know. To find x's value, I will undo the addition. I will subtract 10.25 from both sides of the equation."*

$$x + 10.25 - 10.25 = 106.24 - 10.25$$
$$x = 95.99$$

Through the equation-solving process, x, the number in the equation whose value you didn't know, is 95.99. You obtain this number when you evaluate $106.24 - 10.25$ on the right side of the equation. It shouldn't be a surprise that you get the value of the unknown number, the same way you would do it in arithmetic. You subtract 10.25 from 106.24. So why use an equation? With arithmetic, you have to figure out on your own that you should subtract to find the value of the unknown number. But an equation clues you to what to do to get the answer. By reasoning you must undo the operation that has been done to x, you are guaranteed you will obtain x's correct value.

Replace x with 95.99 in the equation to check it.

Check: $x + 10.25 = 106.24;\ 95.99 + 10.25 \stackrel{?}{=} 106.24; 106.24 \stackrel{\checkmark}{=} 106.24\checkmark$

So 95.99 is the equation's solution. It is a replacement for x that makes the equation $x + 10.25 = 106.24$ a true statement.

Similarly, suppose you have the following problem.

When $\frac{3}{4}$ is subtracted from a number, the difference is $\frac{1}{2}$. What is the number?

Using arithmetic, you should decide to add $\frac{3}{4}$ to $\frac{1}{2}$ to find the unknown number's value. When you do the addition, you get $\frac{1}{2}+\frac{3}{4}=\frac{2}{4}+\frac{3}{4}=\frac{5}{4}$ or $1\frac{1}{4}$.

Here is the way to write the problem using an equation.

Solve $x-\frac{3}{4}=\frac{1}{2}$.

Think: "$\frac{3}{4}$ *is subtracted from x, a number whose value I do not know. To find x's value, I will undo the subtraction. I will add* $\frac{3}{4}$ *to both sides of the equation.*"

$$x-\frac{3}{4}+\frac{3}{4}=\frac{1}{2}+\frac{3}{4}$$
$$x=\frac{2}{4}+\frac{3}{4}$$
$$x=\frac{5}{4} \text{ or } 1\frac{1}{4}$$

Again, you get the value of the unknown number the same way whether you use arithmetic or you write an equation. You add $\frac{3}{4}$ to $\frac{1}{2}$ to obtain $\frac{5}{4}$ or $1\frac{1}{4}$.

Replace x with $\frac{5}{4}$ in the equation to check it.

Check: $x-\frac{3}{4}=\frac{1}{2}$; $\frac{5}{4}-\frac{3}{4}\stackrel{?}{=}\frac{1}{2}$; $\frac{2}{4}\stackrel{?}{=}\frac{1}{2}$; $\frac{\not{2}^{1}}{_{2}\not{4}}\stackrel{?}{=}\frac{1}{2}$; $\frac{1}{2}\stackrel{\checkmark}{=}\frac{1}{2}$ ✓

So $\frac{5}{4}$ is the equation's solution. It is a replacement for x that makes the equation $x-\frac{3}{4}=\frac{1}{2}$ a true statement.

☞Try These

1. Solve and check: $x+\frac{3}{2}=\frac{11}{4}$
2. Solve and check: $x - 7.25 = 3.49$
3. Solve and check: $x + 4.5 = 65.98$
4. Solve and check: $x-\frac{2}{3}=\frac{3}{2}$
5. When 3.24 is added to a number, the sum is 6.23. What is the number?

Solutions

1.
$$x+\frac{3}{2}=\frac{11}{4}$$
$$x+\frac{3}{2}-\frac{3}{2}=\frac{11}{4}-\frac{3}{2}$$
$$x=\frac{11}{4}-\frac{6}{4}$$
$$x=\frac{5}{4}$$

Check: $x+\frac{3}{2}=\frac{11}{4};\ \frac{5}{4}+\frac{3}{2}\stackrel{?}{=}\frac{11}{4};\ \frac{5}{4}+\frac{6}{4}\stackrel{?}{=}\frac{11}{4};\ \frac{11}{4}\stackrel{\checkmark}{=}\frac{11}{4}\checkmark$

2.
$$x-7.25=3.49$$
$$x-7.25+7.25=3.49+7.25$$
$$x=10.74$$

Check: $x-7.25=3.49;\ 10.74-7.25\stackrel{?}{=}3.49;\ 3.49\stackrel{\checkmark}{=}3.49\checkmark$

3.
$$x+4.5=65.98$$
$$x+4.5-4.5=65.98-4.5$$
$$x=61.48$$

Check: $x+4.5=65.98;\ 61.48+4.5\stackrel{?}{=}65.98;\ 65.98\stackrel{\checkmark}{=}65.98\checkmark$

4.
$$x-\frac{2}{3}=\frac{3}{2}$$
$$x-\frac{2}{3}+\frac{2}{3}=\frac{3}{2}+\frac{2}{3}$$
$$x=\frac{9}{6}+\frac{4}{6}$$
$$x=\frac{13}{6}$$

Check: $x-\frac{2}{3}=\frac{3}{2};\ \frac{13}{6}-\frac{2}{3}\stackrel{?}{=}\frac{3}{2};\ \frac{13}{6}-\frac{4}{6}\stackrel{?}{=}\frac{3}{2};\ \frac{9}{6}\stackrel{?}{=}\frac{3}{2};\ \frac{3}{2}\stackrel{\checkmark}{=}\frac{3}{2}\checkmark$

5. Let x = the number. Write and solve the equation: $x + 3.24 = 6.23$.

$$x+3.24=6.23$$
$$x+3.24-3.24=6.23-3.24$$
$$x=2.99$$

The number is 2.99.

Check: $x+3.24=6.23;\ 2.99+3.24\stackrel{?}{=}6.23;\ 6.23\stackrel{\checkmark}{=}6.23\checkmark$

Solving One-Step Mathematical Equations of the Forms $px = q$ and $\frac{x}{p} = q$

Suppose you have the following problem.

When a number is multiplied by 9, the product is 11.25. What is the number?

In arithmetic, to find the value of the number, you must decide to divide 11.25 by 9. When you do the division, you get 11.25 ÷ 9 = 1.25.

Here is the way to write the problem using an equation.

Solve $9x = 11.25$.

Think: *"x, a number whose value I do not know, is multiplied by 9. To find x's value, I will undo the multiplication. I will divide both sides of the equation by 9."*

$$9x = 11.25$$
$$\frac{9x}{9} = \frac{11.25}{9}$$
$$\frac{{}^{1}\cancel{9}x}{\cancel{9}_{1}} = \frac{\cancel{11.25}^{1.25}}{{}_{1}\cancel{9}}$$
$$1x = 1.25$$
$$x = 1.25$$

Tip: Use the method shown in the section "Performing Operations with Multi-Digit Decimals Using Standard Algorithms" in Chapter 2 to divide 11.25 by 9.

$$\begin{array}{r} 1.25 \\ 9\overline{)11.25} \\ -9 \\ \hline 22 \\ -18 \\ \hline 45 \\ -45 \\ \hline 0 \end{array}$$

Through the equation-solving process, x, the number in the equation whose value you didn't know, is 1.25. You obtain this number when you evaluate $\frac{11.25}{9}$ on the right side of the equation. So you get the value of the unknown number the same way whether you use arithmetic or you write an equation. You divide 11.25 by 9.

Replace x with 1.25 in the equation to check it.

Check: $9x = 11.25$; $9(1.25) \stackrel{?}{=} 11.25$; $11.25 \stackrel{\checkmark}{=} 11.25\checkmark$

So 1.25 is the equation's solution. It is a replacement for x that makes the equation $9x = 11.25$ a true statement.

Similarly, suppose you have the following problem.

When a number is divided by 4, the quotient is 14. What is the number?

Using arithmetic, you should multiply 14 by 4 to find the unknown number's value. When you do the multiplication, you get $4 \cdot 14 = 56$.

Here is the way to write the problem using an equation.

Solve $\frac{x}{4} = 14$.

Think: *"x, a number whose value I do not know, is divided by 4. To find x's value, I will undo the division. I will multiply both sides of the equation by 4."*

$$\frac{x}{4} = 14$$
$$4 \cdot \frac{x}{4} = 4 \cdot 14$$
$$\frac{\overset{1}{\cancel{4}}}{1} \cdot \frac{x}{\underset{1}{\cancel{4}}} = 4 \cdot 14$$
$$1x = 56$$
$$x = 56$$

Again, you get the value of the unknown number the same way whether you use arithmetic or you write an equation. You multiply 14 by 4 to obtain 56.

Replace x with 56 in the equation to check it.

Check: $\frac{x}{4} = 14$; $\frac{56}{4} \stackrel{?}{=} 14$; $14 \stackrel{\checkmark}{=} 14 \checkmark$

So 56 is the equation's solution. It is a replacement for x that makes the equation $\frac{x}{4} = 14$ a true statement.

☞ Try These

1. Solve and check: $4.5x = 17.1$
2. Solve and check: $\frac{x}{15} = 6.2$
3. Solve and check: $\frac{1}{4}x = 18.3$
4. Solve and check: $\frac{x}{4} = 18.3$
5. When a number is multiplied by 1.5, the product is 19.2. What is the number?

Solutions

1. $4.5x = 17.1$

$\frac{4.5x}{4.5} = \frac{17.1}{4.5}$

$\frac{{}^{1}\cancel{4.5}x}{{}_{1}\cancel{4.5}} = \frac{\cancel{17.1}^{3.8}}{\cancel{4.5}_{1}}$

$1x = 3.8$

$x = 3.8$

Check: $4.5x = 17.1;\ 4.5(3.8) \stackrel{?}{=} 17.1;\ 17.1 \stackrel{\checkmark}{=} 17.1\checkmark$

2. $\frac{x}{15} = 6.2$

$15 \cdot \frac{x}{15} = 15 \cdot 6.2$

$\frac{\cancel{15}^{1}}{1} \cdot \frac{x}{\cancel{15}_{1}} = 93$

$1x = 93$

$x = 93$

Check: $\frac{x}{15} = 6.2;\ \frac{93}{15} \stackrel{?}{=} 6.2;\ \frac{93.0}{15} \stackrel{?}{=} 6.2;\ 6.2 \stackrel{\checkmark}{=} 6.2\checkmark$

3. $\frac{1}{4}x = 18.3$

$\frac{1}{4}x \div \frac{1}{4} = 18.3 \div \frac{1}{4}$

$\frac{1}{4}x \cdot \frac{4}{1} = \frac{18.3}{1} \cdot \frac{4}{1}$

$\frac{1}{{}_{1}\cancel{4}}x \cdot \frac{\cancel{4}^{1}}{1} = \frac{73.2}{1}$

$1x = 73.2$

$x = 73.2$

Check: $\frac{1}{4}x = 18.3;\ \frac{1}{4} \cdot 73.2 \stackrel{?}{=} 18.3;\ \frac{1}{4} \cdot \frac{73.2}{1} \stackrel{?}{=} 18.3;\ \frac{73.2}{4} \stackrel{?}{=} 18.3;\ 18.3 \stackrel{\checkmark}{=} 18.3\checkmark$

4. $\frac{x}{4} = 18.3$

$4 \cdot \frac{x}{4} = 4 \cdot 18.3$

$\frac{\cancel{4}^{1}}{1} \cdot \frac{x}{\cancel{4}_{1}} = 73.2$

$1x = 73.2$

$x = 73.2$

Check: $\frac{x}{4} = 18.3;\ \frac{73.2}{4} \stackrel{?}{=} 18.3;\ 18.3 \stackrel{\checkmark}{=} 18.3\checkmark$

5. Let x = the number. Write and solve the equation: $1.5x = 19.2$.

$$1.5x = 19.2$$
$$\frac{1.5x}{1.5} = \frac{19.2}{1.5}$$
$$\frac{\cancel{1.5}^{1}x}{_{1}\cancel{1.5}} = \frac{\cancel{19.2}^{12.8}}{\cancel{1.5}_{1}}$$
$$1x = 12.8$$
$$x = 12.8$$

The number is 12.8.

Check: $1.5x = 19.2$; $1.5(12.8) \stackrel{?}{=} 19.2$; $19.2 \stackrel{\checkmark}{=} 19.2$ ✓

Solving Real-World One-Step Equations of the Forms $x + p = q$, $x - p = q$, $px = q$, and $\frac{x}{p} = q$

The following table summarizes solving mathematical one-step, single-variable equations.

Type	How to Solve It	Example	How to Solve It	How to Check It
$x + p = q$	Subtract the number p from both sides of the equation.	$x + 12 = 30$	Subtract 12 from both sides. $x+12=30$ $x+12-12=30-12$ $x=18$	$x+12=30$ $18+12\stackrel{?}{=}30$ $30\stackrel{\checkmark}{=}30$
$x - p = q$	Add the number p to both sides of the equation.	$x - 13 = 25$	Add 13 to both sides. $x-13=25$ $x-13+13=25+13$ $x=38$	$x-13=25$ $38-13\stackrel{?}{=}25$ $25\stackrel{\checkmark}{=}25$
$px = q$	Divide both sides of the equation by the number p.	$2x = 24$	Divide both sides by 2. $2x=24$ $\frac{2x}{2}=\frac{24}{2}$ $\frac{\cancel{2}x}{\cancel{2}}=\frac{\cancel{24}^{12}}{\cancel{2}}$ $x=12$	$2x=24$ $2\cdot12\stackrel{?}{=}24$ $24\stackrel{\checkmark}{=}24$
$\frac{x}{p} = q$	Multiply both sides of the equation by the number p.	$\frac{x}{3}=65$	Multiply both sides by 3: $\frac{x}{3}=65$ $3\cdot\frac{x}{3}=3\cdot65$ $\frac{\cancel{3}^{1}}{1}\cdot\frac{x}{_{1}\cancel{3}}=195$ $x=195$	$\frac{x}{3}=65$ $\frac{195}{3}\stackrel{?}{=}65$ $65\stackrel{\checkmark}{=}65$

Many real-world problems can be modeled and solved using one-step, single-variable equations. To solve such problems, read the problem carefully. Look for a sentence that contains words or phrases such as "what is," "what was," "find," "how many," and "determine" to help you identify what you are to find. Let the variable represent this unknown quantity. (***Tip:*** Be precise in specifying a variable. State its units, if any.) Then write and solve an equation that represents the facts given in the problem. Here is an example.

> A customer gets a \$15 discount off the original price of a shirt. The price of the shirt after the discount is \$58.96. What was the shirt's original price?

Think: *"I need to find the shirt's original price. The discount decreased the shirt's original price by \$15. The shirt's price after the discount is \$58.96."*

Let s = the shirt's original price in dollars. Write and solve an equation that represents the facts given in the problem.

$$s - \$15 = \$58.96$$
$$s - \$15 + \$15 = \$58.96 + \$15$$
$$s = \$73.96$$

The shirt's original price was \$73.96.

Check: The shirt's original price was \$73.96. After \$15 is taken off, the price is \$73.96 – \$15 = \$58.96.✓

Tip: When you solve a real-world problem, check your answer by making sure it is consistent with the facts given in the problem. This way of checking helps you feel confident your equation is written correctly.

☞ Try These

1. The original price of a watch is increased by \$16.92. The price of the watch after the increase is \$59.22. Find the original price of the watch.
2. A box contains red and black marbles only. The number of red marbles in the box is 12 less than the number of black marbles in the box. If there are 15 red marbles in the box, how many black marbles are in the box?
3. The distance to City B from Indio's house is 1.5 times the distance from his house to City A. If Indio's house is 45 miles from City B, what is the distance from his house to City A?
4. Three boys equally share the cost of a pizza. If each boy paid \$6.25, what was the cost of the pizza?

Solutions

1. Let c = the original price of the watch in dollars. Write and solve an equation that represents the facts given in the problem.

$$c + \$16.92 = \$59.22$$
$$c + \$16.92 - 16.92 = \$59.22 - 16.92$$
$$c = \$42.30$$

The watch's original price is \$42.30.

Check: The watch's original price is \$42.30. After an increase of \$16.92, the price of the watch is \$42.30 + \$16.92 = \$59.22.✓

2. Let b = the number of black marbles in the box. Write and solve an equation that represents the facts given in the problem.

$$\begin{aligned} b-12&=15 \\ b-12+12&=15+12 \\ b&=27 \end{aligned}$$

There are 27 black marbles in the box.

Check: There are 27 black marbles in the box. The number of red marbles in the box is 12 less than 27, which is 27 – 12 = 15.✓

3. Let a = the distance from Indio's house to City A in miles. Write and solve an equation that represents the facts given in the problem.

$$\begin{aligned} 1.5a&=45\text{ miles} \\ \frac{1.5a}{1.5}&=\frac{45\text{ miles}}{1.5} \\ \frac{{}^{1}\cancel{1.5}a}{{}_{1}\cancel{1.5}}&=\frac{\cancel{45}^{30}\text{ miles}}{\cancel{1.5}_{1}} \\ a&=30\text{ miles} \end{aligned}$$

The distance from Indio's house to City A is 30 miles.

Check: The distance from Indio's house to City A is 30 miles. The distance to City B from Indio's house is 1.5 times 30 miles, which is 45 miles.✓

4. Let c = the cost of the pizza in dollars. Write and solve an equation that represents the facts given in the problem.

$$\begin{aligned} \frac{c}{3}&=\$6.25 \\ 3\cdot\frac{c}{3}&=3\cdot\$6.25 \\ \frac{\cancel{3}^{1}}{1}\cdot\frac{c}{{}_{1}\cancel{3}}&=\$18.75 \\ c&=\$18.75 \end{aligned}$$

The cost of the pizza was \$18.75.

Check: The cost of the pizza was \$18.75. The cost equally shared by three boys is $\frac{\$18.75}{3}$, which is \$6.25.✓

Understanding and Using Inequalities of the Form $x < c$ and $x > c$

(CCSS.Math.Content.6.EE.B.8)

In the inequality $x < c$, the variable x represents all numbers less than c. For instance, the solution to the inequality $x < 8$ is all numbers less than 8. Any number less than 8 makes the inequality a true statement. On the number line, the solution is all numbers to the left of 8, as shown below. An open circle at 8 indicates 8 is NOT included in the solution. The heavy line to the left of 8 represents all numbers less than 8. Any number that lies to the left of 8 on the number line makes the inequality $x < 8$ a true statement.

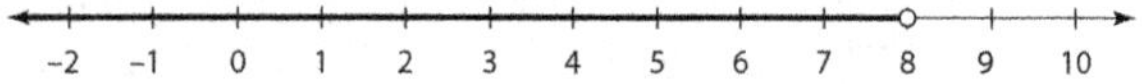

Similarly, in the inequality $x > c$, the variable x represents all numbers greater than c. For example, the solution to the inequality $x > 2$ is all numbers greater than 2. Any number greater than 2 makes the inequality a true statement. On the number line, the solution is all numbers to the right of 2, as shown below. An open circle at 2 indicates 2 is NOT included in the solution. The heavy line to the right of 2 represents all numbers greater than 2. Any number that lies to the right of 2 on the number line makes the inequality $x > 2$ a true statement.

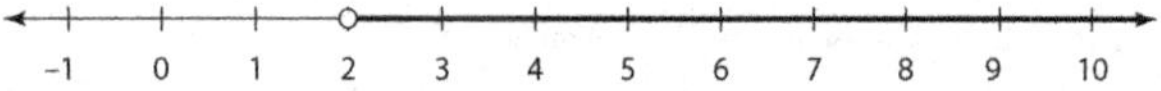

You can combine the inequalities $x < b$ and $x > a$, where $a < b$, into the double inequality $a < x < b$. This inequality means x represents all numbers between a and b. For instance, the solution to $2 < x < 8$ is all numbers between 2 and 8. The graph of the solution is a line segment between 2 and 8, but not including the endpoints 2 and 8, as shown here.

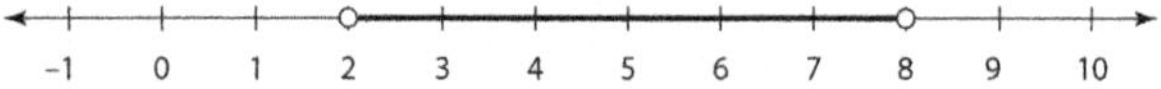

Use inequalities to represent a constraint or condition in a real-world problem. Here is an example.

> A high school coach spent less than \$230 to buy new basketballs for the coming year. Write an inequality to represent this situation. Then graph the solution.

The amount the coach spent is between \$0 and \$230. Let b = the amount of money in dollars that the coach spent. Then, the inequality $\$0 < b < \$230$ represents the given situation.

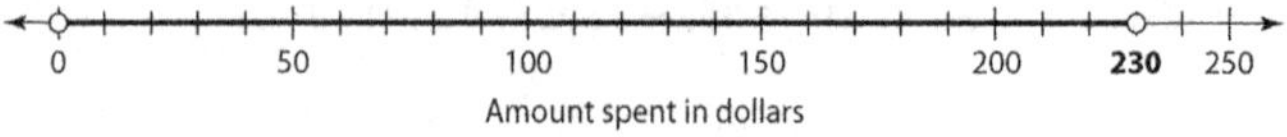

Try These

1. Tell which number or numbers from the set make the inequality a true statement.

   (a) $x < 8$: $\left\{-1{,}000,\ 0,\ \frac{31}{4},\ 3.14,\ 7.99,\ 8,\ 8.01,\ 5{,}900\right\}$

   (b) $x > 2$: $\left\{-10,\ 0,\ \frac{3}{4},\ 2,\ 2.001,\ 3.14,\ \frac{7}{2},\ 6{,}000\right\}$

   (c) $2 < x < 8$: $\left\{-1{,}000,\ 0,\ 2,\ 2.001,\ \frac{31}{4},\ 8,\ 8.01,\ 5{,}900\right\}$

2. Todd spent less than $100 for school supplies.

   (a) Write an inequality to represent the situation.
   (b) Graph the solution of the inequality.

3. Daisuke worked more than 40 hours on his science project.

   (a) Write an inequality to represent the situation.
   (b) Graph the solution of the inequality.

4. Ramy lives in a town that is less than 300 feet above sea level.

   (a) Write an inequality to represent the situation.
   (b) Graph the solution of the inequality.

Solutions

1. **(a)** $-1{,}000,\ 0,\ \frac{31}{4},\ 3.14,\ 7.99$

   (b) $2.001,\ 3.14,\ \frac{7}{2},\ 6{,}000$

   (c) $2.001,\ \frac{31}{4}$

2. **(a)** Let m = the amount of money in dollars that Todd spent. Then, the inequality $\$0 < m < \$100$ represents the given situation.

   (b)

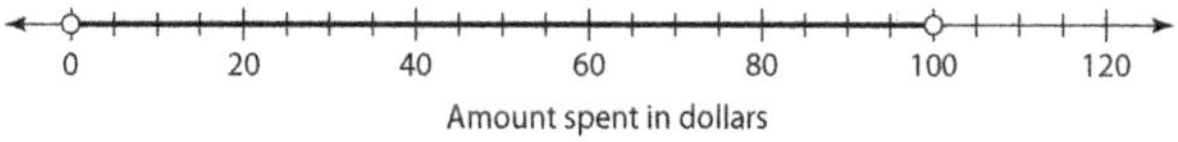

3. **(a)** Let h = the number of hours Daisuke worked. Then, $h > 40$ hours represents the given situation.

   (b)

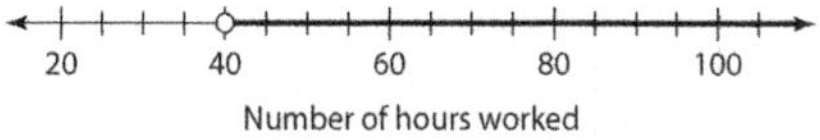

4. **(a)** Let a = the altitude of the town. Then, $a < 300$ feet represents the given situation.

   (b)

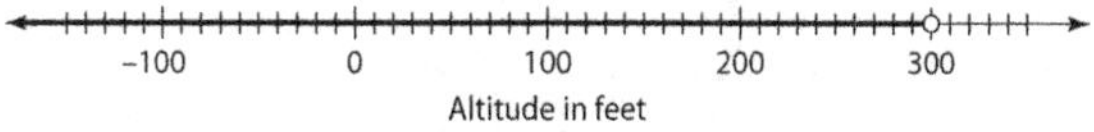

Representing and Analyzing Relationships Between Two Variables

(CCSS.Math.Content.6.EE.C.9)

Sara's parents have a membership to a fitness center. For \$5, Sara can obtain a 1-day swim pass for the pool at the fitness center. For each guest she brings along, the cost for the guest's 1-day swim pass is \$3. No more than five guests per day are allowed. Write an equation to represent the total cost for 1-day swim passes for Sara and her guests.

The total cost for 1-day swim passes is \$5 for Sara and \$3 for each guest. Let n represent the number of guests and c represent the total cost in dollars. Then the equation $c = \$5 + \$3n$ represents the total cost for 1-day swim passes for Sara and n guests.

Create a table to show the total cost for 1-day swim passes when Sara brings zero to five guests.

Number of Guests (*n*)	**Total Cost in Dollars (*c*)**
0	5
1	8
2	11
3	14
4	17
5	20

In the equation $c = \$5 + \$3n$, n is the **independent variable** and c is the **dependent variable.** The total cost, c, depends on the number of guests, n. The value of the dependent variable, c, depends on the value of the independent variable, n. The table shows that as the number of guests increases, the total cost increases.

Graph the ordered pairs (0, 5), (1, 8), (2, 11), (3, 14), (4, 17), and (5, 20) from the table to see a visual depiction of the relationship between n and c. Graph the independent variable on the horizontal axis, and the dependent variable on the vertical axis. ***Tip:*** Do not connect the points because the number of guests, n, must be a whole number.

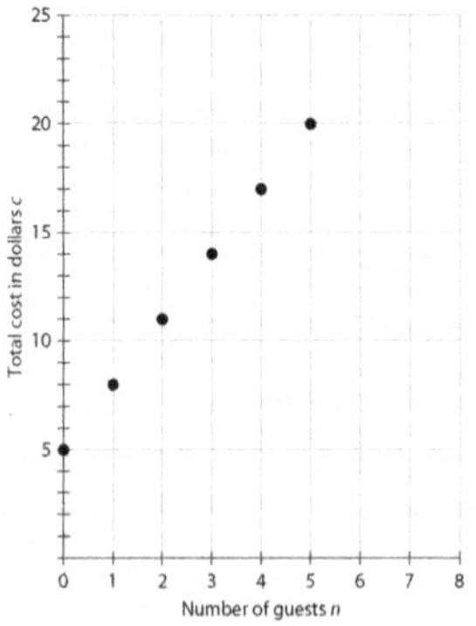

Try These

1. Ramy subscribes to an on-demand Internet streaming movie provider service. The streaming service costs $8.65 per month for access to its library of movie titles. For other movie titles, such as movies recently released in theaters, there is a charge of $9.99 each.
   - **(a)** Write an equation to model the situation.
   - **(b)** Create a table to show the total cost per month, given that Ramy might purchase from 2 to 7 movies in a month.
   - **(c)** Identify the independent and dependent variables.
   - **(d)** Create a graph that depicts the equation in part (a).
2. Brandi makes a trip driving an average speed of 65 miles per hour.
   - **(a)** Write an equation to model the relationship between time traveled and distance traveled.
   - **(b)** Create a table to show the number of miles traveled for 1 to 4 hours.
   - **(c)** Identify the independent and dependent variables.
   - **(d)** Create a graph that depicts the equation in part (a).

Solutions

1. **(a)** Let n represent the number of movies purchased in a month and c represent the total monthly cost in dollars. Then the equation $c = \$8.65 + \$9.99n$ represents the total monthly cost.

(b)

Number of Movies Purchased (*n*)	Total Monthly Cost in Dollars (*c*)
2	28.63
3	38.62
4	48.61
5	58.60
6	68.59
7	78.58

(c) The total monthly cost, c, depends on the number, n, of movies purchased. The independent variable is n, the number of movies purchased, and the dependent variable is c, the total monthly cost.

(d)

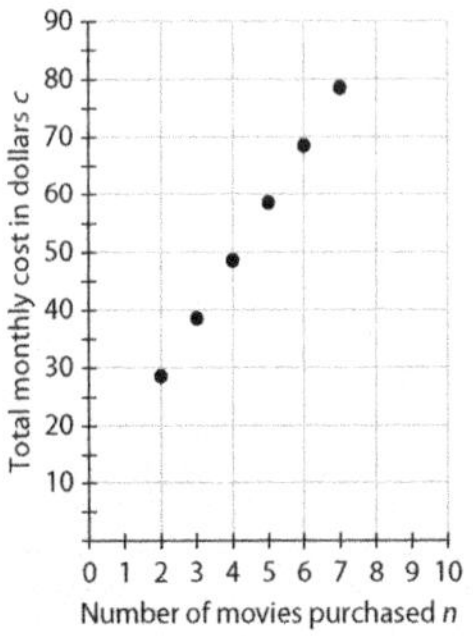

2. **(a)** Let t represent the time traveled in hours and d represent the distance traveled in miles. Then the equation $d = 65t$ represents the distance in miles traveled in t hours.

(b)

Time in Hours (t)	**Distance in Miles (d)**
1	65
2	130
3	195
4	260

(c) The distance, c, depends on the time, t, traveled. The independent variable is t, the time in hours, and the dependent variable is d, the distance in miles.

(d)

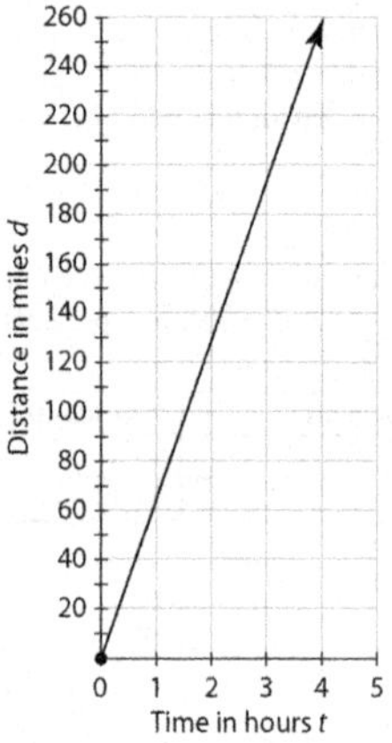

4. Geometry

In this chapter, you will determine the area of triangles, quadrilaterals, and other polygonal regions. You will find the volume of right rectangular prisms. You will use your knowledge of the coordinate plane to find the dimensions of polygons given coordinates of their vertices. You will use nets to find the surface area of solid figures. In addition, you will apply geometric concepts to solve mathematical and real-world problems.

Finding the Area of Triangles

(CCSS.Math.Content.6.G.A.1)

A useful way to classify triangles is according to their interior angles. An **acute** triangle has three acute angles. A **right** triangle has exactly one right angle. An **obtuse** triangle has exactly one obtuse angle. Here are examples.

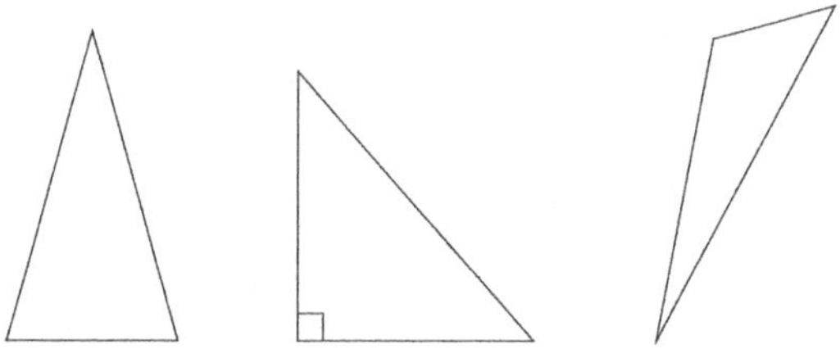

Acute triangle   Right triangle   Obtuse triangle

To find the area of a triangle, you must know the measures of the triangle's **base** and **height.** The base can be any of the triangle's three sides. The height for that base is the length of the altitude to that base. The **altitude** is a perpendicular line segment from the opposite vertex to the line containing the base.

The formula for the area of a triangle is $A = \frac{1}{2}bh$, where b is the length of a base of the triangle and h is the height for that base. For every triangle, you have three choices for the base. Each choice will have a corresponding height. When you are finding the area of a triangle, pick any convenient side of the triangle to serve as the base in the formula.

In an acute triangle, the altitudes lie within the triangle. Here is an example.

Determine the area of the given acute triangle.

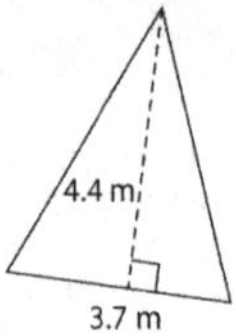

As the base, choose the side that has an altitude drawn to it. Then, $b = 3.7$ m and $h = 4.4$ m. Plugging into the formula yields

$$A = \frac{1}{2}bh = \frac{1}{2}(3.7\text{ m})(4.4\text{ m}) = \frac{(3.7\text{ m})(4.4\text{ m})}{2} = \frac{16.28\text{ m}^2}{2} = 8.14\text{ m}^2$$

The area of the triangle is 8.14 m^2.

Tip: Recall that unit times unit equals unit². So (m)(m) = m², (ft)(ft) = ft², (cm)(cm) = cm², (in)(in) = in², and so forth. Regardless of the shape of the figure, area units are always square units.

In a right triangle, the two legs meet at a right angle. When it is convenient to do so, you can designate one leg as the base and the other as the altitude. (***Tip:*** It doesn't matter which is designated the base and which is designated the altitude.) Here is an example.

Determine the area of the given right triangle.

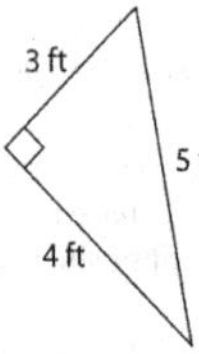

Designate the shorter leg the base, and the other leg, the altitude. Then, $b = 3$ ft and $h = 4$ ft. Plugging into the formula yields

$$A = \frac{1}{2}bh = \frac{1}{2}(3\text{ ft})(4\text{ ft}) = \frac{(3\text{ ft})(\cancel{4}^2\text{ ft})}{{}_1\cancel{2}} = \frac{6\text{ ft}^2}{1} = 6\text{ ft}^2$$

The area of the triangle is 6 ft^2.

In an obtuse triangle, two of the altitudes lie outside the triangle. Here is an example.

Determine the area of the given obtuse triangle.

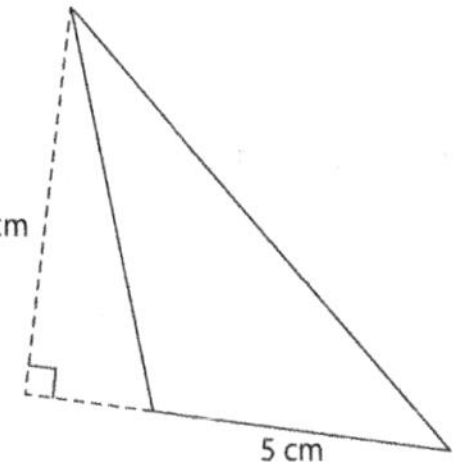

As the base, choose the side whose corresponding height is given. Let $b = 5$ cm and $h = 6$ cm. Plugging into the formula yields

$$A = \frac{1}{2}bh = \frac{1}{2}(5\text{ cm})(6\text{ cm}) = \frac{(5\text{ cm})(\cancel{6}^{3}\text{ cm})}{{}_{1}\cancel{2}} = \frac{15\text{ cm}^2}{1} = 15\text{ cm}^2$$

The area of the triangle is 15 cm^2.

☞ Try These

1. Determine the area of the given triangle.

(a)

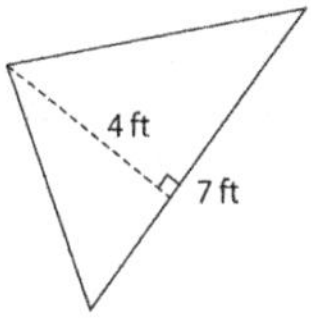

(b)

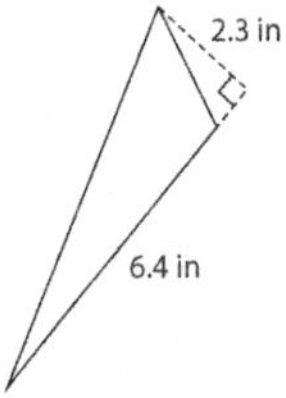

(c)

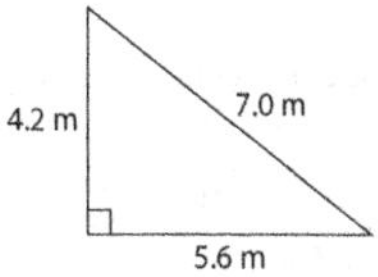

2. Donna and Jeremy have a vegetable garden in the shape of a right triangle. The garden measures 8 feet by 15 feet by 17 feet. How many square feet of area do Donna and Jeremy have for their vegetable garden?

Solutions

1. **(a)** $A=\frac{1}{2}bh=\frac{1}{2}(4\text{ ft})(7\text{ ft})=\frac{(\,^{2}\cancel{4}\text{ ft})(7\text{ ft})}{_{1}\cancel{2}}=\frac{14\text{ ft}^2}{1}=14\text{ ft}^2$

   (b) $A=\frac{1}{2}bh=\frac{1}{2}(2.3\text{ in})(6.4\text{ in})=\frac{(2.3\text{ in})(6.4\text{ in})}{2}=\frac{14.72\text{ in}^2}{2}=7.36\text{ in}^2$

   (c) $A=\frac{1}{2}bh=\frac{1}{2}(4.2\text{ m})(5.6\text{ m})=\frac{(4.2\text{ m})(5.6\text{ m})}{2}=\frac{23.52\text{ m}^2}{2}=11.76\text{ m}^2$

2. $A=\frac{1}{2}bh=\frac{1}{2}(8\text{ ft})(15\text{ ft})=\frac{(\,^{4}\cancel{8}\text{ ft})(15\text{ ft})}{_{1}\cancel{2}}=\frac{60\text{ ft}^2}{1}=60\text{ ft}^2$

Finding the Area of Parallelograms and Rectangles

(CCSS.Math.Content.6.G.A.1)

A **parallelogram** is a quadrilateral that has opposite sides parallel and congruent. (***Tip:*** **Congruent** figures have the same size and shape.) A **rectangle** is a parallelogram that has four interior right angles.

The formula for the area of a parallelogram is $A = bh$, where b is the length of one of the parallelogram's sides and h is the perpendicular line segment from one of the opposite vertices to the line containing the base. For every parallelogram, you have four choices for the base. Each choice will have a corresponding height. When you are finding the area of a parallelogram, pick any convenient side of the parallelogram to serve as the base in the formula. Here is an example.

Determine the area of the parallelogram below.

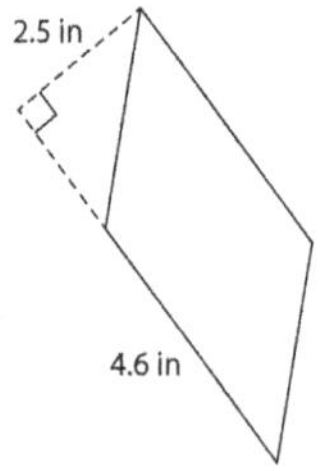

As the base, choose the side whose corresponding height is given. Let $b = 4.6$ in and $h = 2.5$ in. Plugging into the formula yields

$$A = bh = (4.6\text{ in})(2.5\text{ in}) = 11.5\text{ in}^2$$

The area of the parallelogram is 11.5 in^2.

A rectangle is a parallelogram, so you can use the formula $A = bh$ to find its area, the same as you would for any parallelogram. The area of a rectangle is the same as the area of a parallelogram, with the same base and height measurements. Because the length and width of a rectangle are perpendicular, the formula $A = lw$, where l is the rectangle's length and w is its width, yields the same area as the formula $A = bh$. You may use either formula to calculate the area of a rectangle. Here is an example.

Determine the area of the given rectangle.

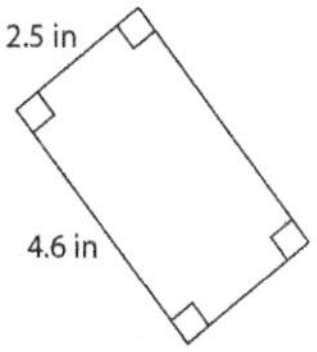

$$A = lw = (4.6\text{ in})(2.5\text{ in}) = 11.5\text{ in}^2$$

The area of the rectangle is 11.5 in^2.

☞ Try These

1. Determine the area of the given figure.

(a)

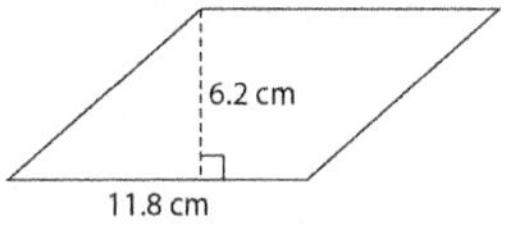

(b)

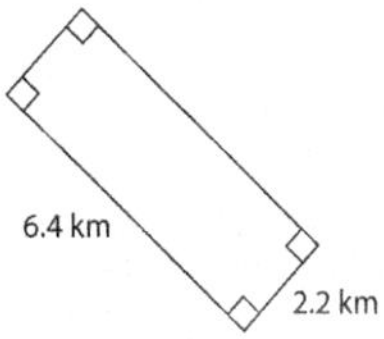

(c)

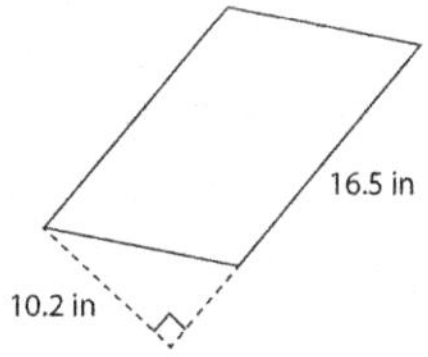

2. Rainey wants to put blue wallpaper on three of the walls in her bedroom and paint the other wall a light blue. The wall she intends to paint is 9 feet high and 10 feet wide and has no door or window. She can purchase the paint in quart containers. One quart of paint will cover 100 square feet of surface. According to the salesperson at the paint store, Rainey will need to apply only one coat of the light blue paint she selected. How many quarts of paint should Rainey purchase? Justify your answer.

Solutions

1. **(a)** $A = (6.2 \text{ cm})(11.8 \text{ cm}) = 73.16 \text{ cm}^2$
   (b) $A = (6.4 \text{ km})(2.2 \text{ km}) = 14.08 \text{ km}^2$
   (c) $A = (10.2 \text{ in})(16.5 \text{ in}) = 168.3 \text{ in}^2$
2. The area of the wall is (9 ft)(10 ft) = 90 square feet. Given one quart of paint covers 100 square feet, Rainey should purchase one quart of the light blue paint.

Finding the Area of Special Quadrilaterals and Other Polygons by Composing or Decomposing into Rectangles, Triangles, and Other Shapes

(CCSS.Math.Content.6.G.A.1)

A **polygon** is a closed plane figure composed of **sides** that are straight line segments. The point at which the two sides of a polygon intersect is a **vertex.** Here are examples of polygons.

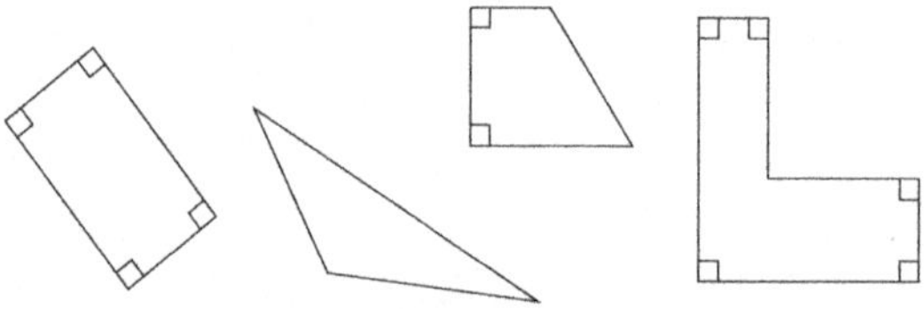

A polygon's area is the area of the region enclosed by its boundary. When you don't know a formula that fits the shape of a polygon, compose or decompose the polygon into right triangles and/or rectangles or other polygons whose areas you can calculate. Find the areas of the shapes whose formulas you know, and then use that information to determine the area of the polygon. Here is an example.

Determine the area of the given trapezoid. ***Tip:*** In this study guide, a **trapezoid** is defined as a quadrilateral that has at least one pair of parallel sides.

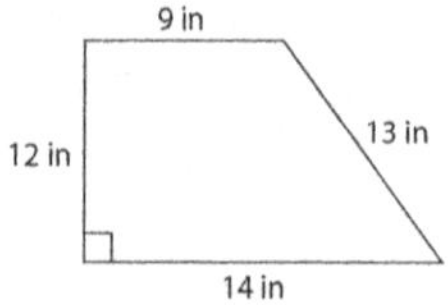

You can find the area of this trapezoid in at least two ways.

Method 1: Decompose the trapezoid into a rectangle and a right triangle as shown below. Determine and label the lengths of the sides of the newly created shapes.

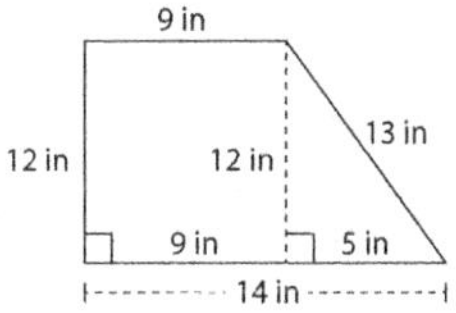

The area, A, of the trapezoid is the sum of the areas of a 9- by 12-inch rectangle and a right triangle with legs 12 inches and 5 inches. Thus,

$$A = (9\text{ in})(12\text{ in}) + \frac{1}{2}(5\text{ in})(12\text{ in}) = 108\text{ in}^2 + 30\text{ in}^2 = 138\text{ in}^2$$

Method 2: Compose the trapezoid with a right triangle to make a rectangle as shown below. Determine and label the lengths of the sides of the newly created shape.

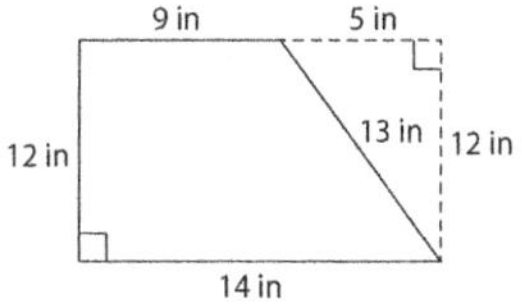

The area, A, of the trapezoid is the difference between the area of a 14- by 12-inch rectangle and a right triangle with legs 12 inches and 5 inches. Thus,

$$A = (14\text{ in})(12\text{ in}) - \frac{1}{2}(5\text{ in})(12\text{ in}) = 168\text{ in}^2 - 30\text{ in}^2 = 138\text{ in}^2$$

☞ Try These

1. Determine the area of the given figure. ***Tip:*** There may be more than one way to find the area.

   (a)

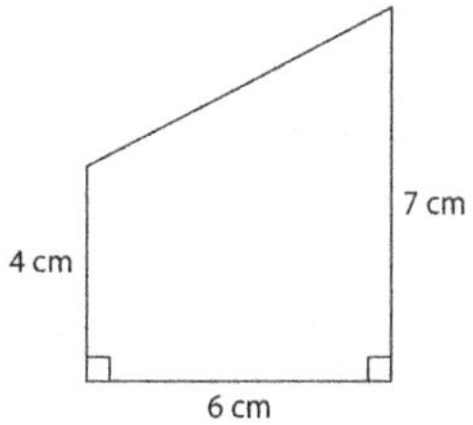

(b)

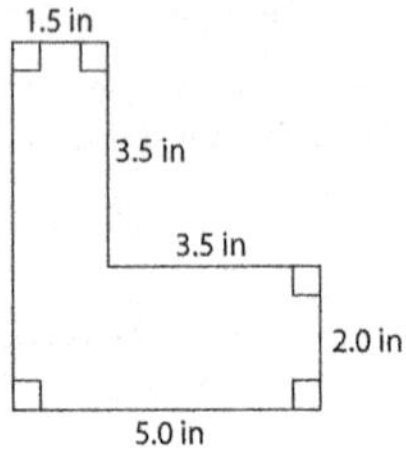

(c)

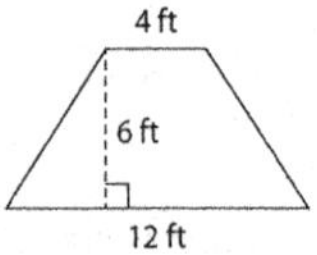

2. A rectangular swimming pool has dimensions 25 meters by 15 meters. The pool is surrounded by a walkway that is 2 meters wide. Find the area of the walkway around the pool.

Solutions

1. **(a)** Cut the figure into a right triangle and a rectangle as shown below.

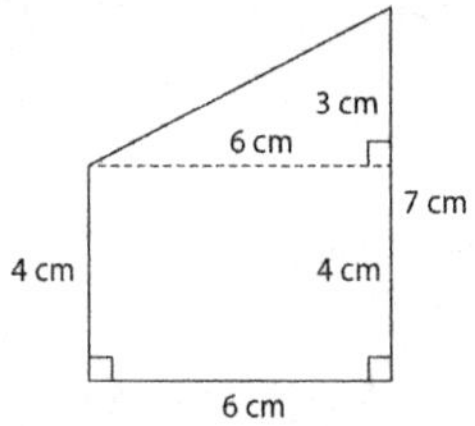

$$A = \frac{1}{2}(6\text{ cm})(3\text{ cm}) + (6\text{ cm})(4\text{ cm}) = 9\text{ cm}^2 + 24\text{ cm}^2 = 33\text{ cm}^2$$

(b) Cut the figure into two rectangles as shown below.

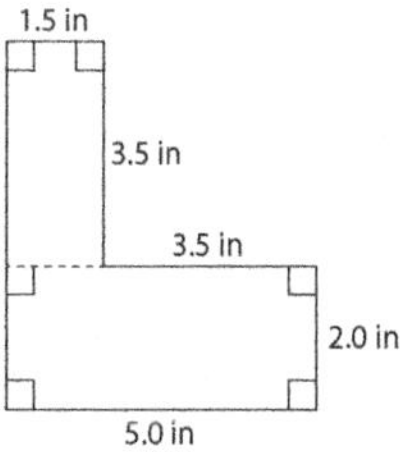

A = (1.5 in)(3.5 in) + (5.0 in)(2.0 in) = 5.25 in^2 + 10 in^2 = 15.25 in^2

(c) Cut the figure into two right triangles and a rectangle as shown below.

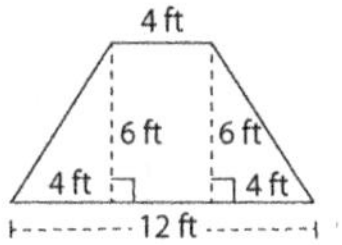

$$A = 2\left(\frac{1}{2}\right)(4\text{ ft})(6\text{ ft}) + (4\text{ ft})(6\text{ ft}) = 24\text{ ft}^2 + 24\text{ ft}^2 = 48\text{ ft}^2$$

2. Make a sketch.

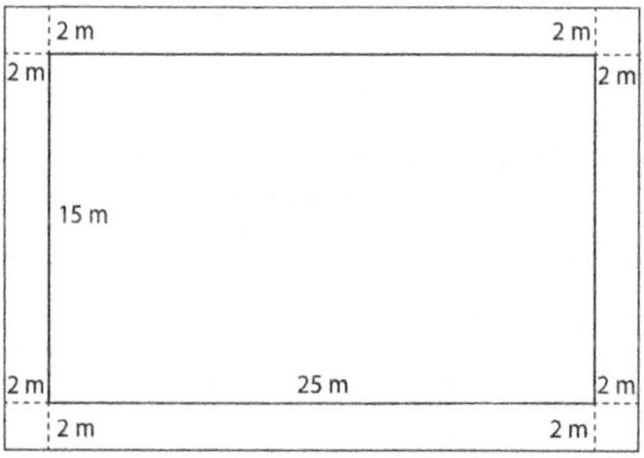

The area of the walkway is the difference between the area of the larger rectangle and the area of the smaller 25 m by 15 m rectangle. The length of the larger rectangle is 25 m + 4 m = 29 m. Its width is 15 m + 4 m = 19 m. The area of the walkway is (29 m)(19 m) – (25 m)(15 m) = 551 m^2 – 375 m^2 = 176 m^2.

Finding Perimeter and Area of Polygons in the Coordinate Plane

(CCSS.Math.Content.6.G.A.3)

The figure shown has vertices A (5.5, 2.2), B (9.1, 2.2), C (9.1, –6.4), D (7.3, –6.4), E (7.3, –2.5), and F (5.5, –2.5).

Find its perimeter and area.

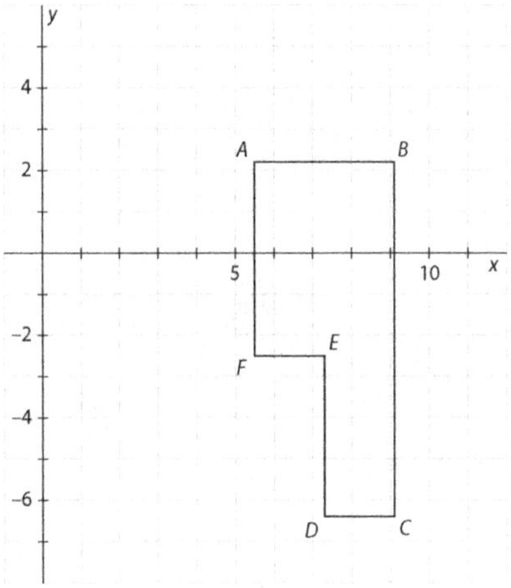

The perimeter, P, of the figure is the sum of the lengths of its six sides. That is, $P = AB + BC + DC + ED + FE + AF$. The line segments composing the figure are either horizontal or vertical lines.

$\overline{AB}$ is a horizontal line. Both of its endpoints, A and B, are in the same quadrant. So $AB = |9.1| - |5.5| = 9.1 - 5.5 = 3.6$ units.

$\overline{BC}$ is a vertical line. Its endpoints, B and C, are in different quadrants. So $BC = |2.2| + |-6.4| = 2.2 + 6.4 = 8.6$ units.

$\overline{DC}$ is a horizontal line. Both of its endpoints, C and D, are in the same quadrant. So $DC = |9.1| - |7.3| = 9.1 - 7.3 = 1.8$ units.

$\overline{ED}$ is a vertical line. Its endpoints, D and E, are in the same quadrant. So $ED = |-6.4| - |-2.5| = 6.4 - 2.5 = 3.9$ units.

$\overline{FE}$ is a horizontal line. Both of its endpoints, E and F, are in the same quadrant. So $FE = |7.3| - |5.5| = 7.3 - 5.5 = 1.8$ units.

$\overline{AF}$ is a vertical line. Its endpoints, A and F, are in different quadrants. So $AF = |2.2| + |-2.5| = 2.2 + 2.5 =$ 4.7 units.

$$\begin{aligned} P &= AB + BC + DC + ED + FE + AF \\ &= 3.6 \text{ units} + 8.6 \text{ units} + 1.8 \text{ units} + 3.9 \text{ units} + 1.8 \text{ units} + 4.7 \text{ units} \\ &= 24.4 \text{ units} \end{aligned}$$

Tip: See "Finding Distances Between Points in the Coordinate Plane" in Chapter 2 for guidance on determining the lengths of horizontal and vertical lines in the coordinate plane.

To compute the figure's area, cut it into two rectangles as shown below.

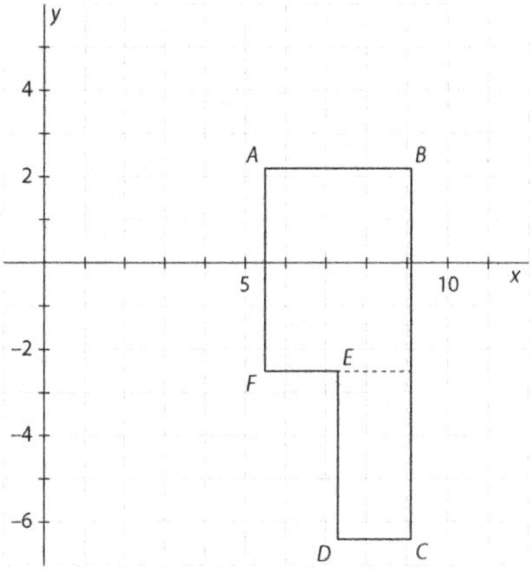

The figure's area, A, equals the sum of the areas of the two rectangles:

$$\begin{aligned} A &= (AB)(AF) + (DC)(ED) \\ &= (3.6 \text{ units})(4.7 \text{ units}) + (1.8 \text{ units})(3.9 \text{ units}) \\ &= 16.92 \text{ units}^2 + 7.02 \text{ units}^2 \\ &= 23.94 \text{ units}^2 \end{aligned}$$

Here's another example.

The vertices of the triangle shown are $A(-3, 3)$, $B(1, 5)$, and $C(4, 2)$. Determine the area of triangle ABC.

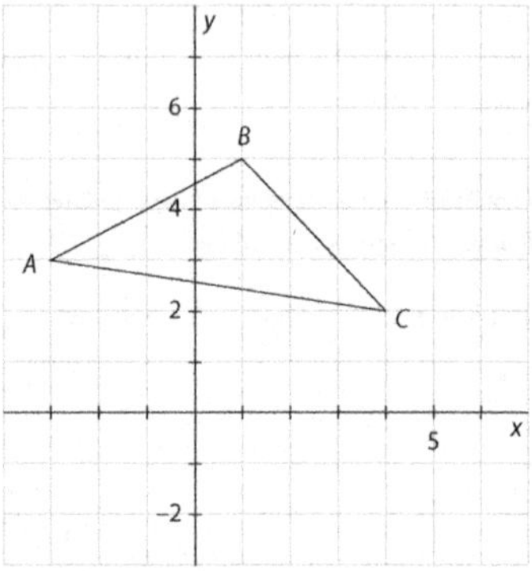

The figure has no horizontal or vertical sides, so you cannot find the lengths of the sides by the methods in Chapter 2. To determine the area, enclose the triangle in a rectangle as shown below. Make the rectangle's top side parallel to the x-axis and passing through vertex B of the triangle. Make the rectangle's bottom side parallel to the x-axis and passing through vertex C of the triangle. Make the left side of the rectangle perpendicular to its top and bottom sides and passing through vertex A. Make the right side of the rectangle perpendicular to its top and bottom sides and passing through vertex C. To help keep track of your work, label the coordinates of the vertices in the figure.

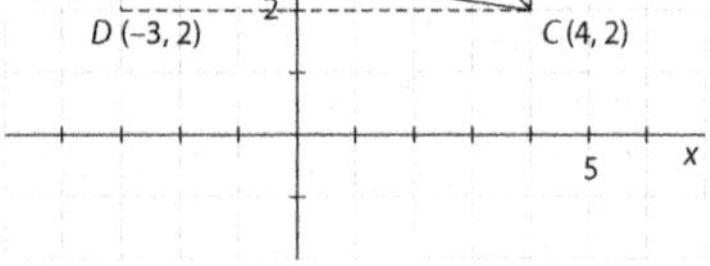

The area, A, of triangle ABC is the area of rectangle $EDCF$ minus the areas of right triangles ADC, CFB, and BEA. In this figure, you can find the lengths of the sides of the figures by counting.

Rectangle $EDCF$ has dimensions 7 by 3 units. Triangle ADC has base 7 units and height 1 unit. Triangle CFB has base 3 units and height 3 units. And triangle BEA has base 4 units and height 2 units.

$$A = (7 \text{ units})(3 \text{ units}) - \frac{1}{2}(7 \text{ units})(1 \text{ unit}) - \frac{1}{2}(3 \text{ units})(3 \text{ units}) - \frac{1}{2}(4 \text{ units})(2 \text{ units})$$
$$= 21 \text{ units}^2 - 3.5 \text{ units}^2 - 4.5 \text{ units}^2 - 4 \text{ units}^2$$
$$= 9 \text{ units}^2$$

Triangle ABC has area 9 units^2.

☞ Try These

1. The figure shown has vertices A (5, 2), B (9, 2), C (9, –6), D (7, –6), E (7, –2), and F (5, –2).

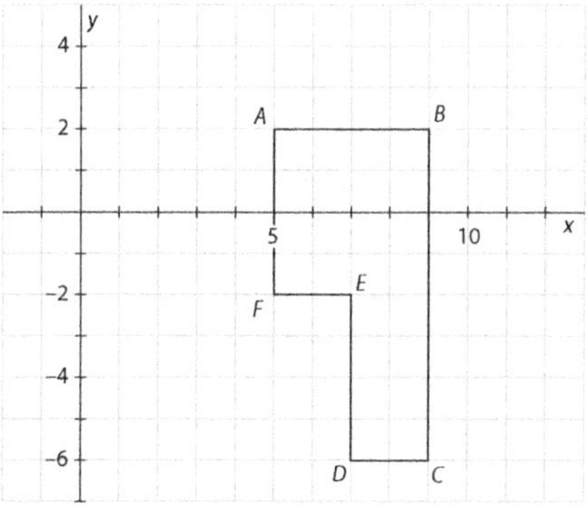

(a) Find the perimeter of the figure.
(b) Find the area of the figure.

2. The triangle shown has vertices X (1, 0), Y (3, –2), and Z (4, 5). Determine its area.

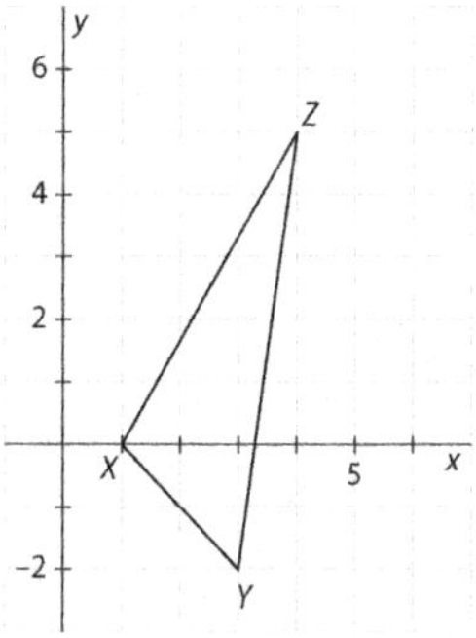

3. Find the area of the figure shown.

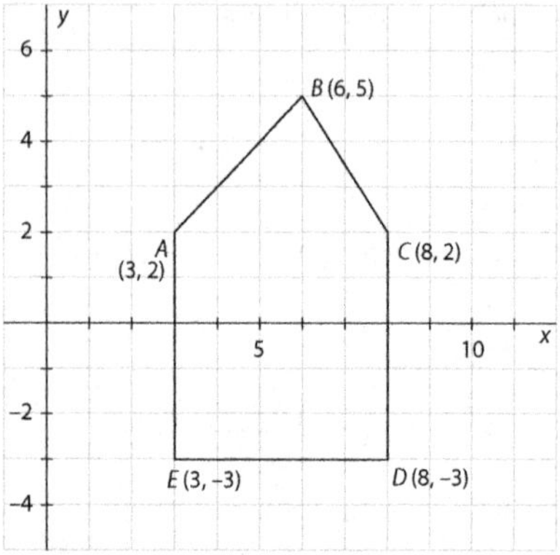

4. The grid shows Tyson's running path through his neighborhood. If each grid line represents 10 meters, how many meters does Tyson run one time around this path?

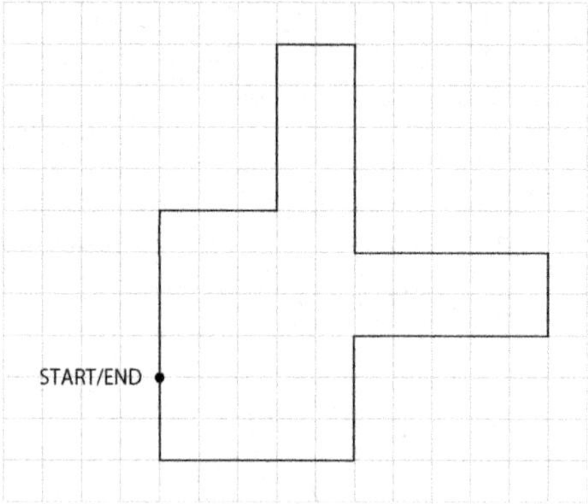

Solutions

1. **(a)** $P = AB + BC + DC + ED + FE + AF$

$AB = |9| - |5| = 9 - 5 = 4$ units

$BC = |2| + |-6| = 2 + 6 = 8$ units

$DC = |9| - |7| = 9 - 7 = 2$ units

$ED = |-6| - |-2| = 6 - 2 = 4$ units

$FE = |7| - |5| = 7 - 5 = 2$ units

$AF = |2| + |-2| = 2 + 2 = 4$ units

$$\begin{aligned} P &= AB + BC + DC + ED + FE + AF \\ &= 4 \text{ units} + 8 \text{ units} + 2 \text{ units} + 4 \text{ units} + 2 \text{ units} + 4 \text{ units} \\ &= 24 \text{ units} \end{aligned}$$

(b) To compute the figure's area, cut it into a 4-unit by 4-unit square and a 2-unit by 4-unit rectangle as shown below.

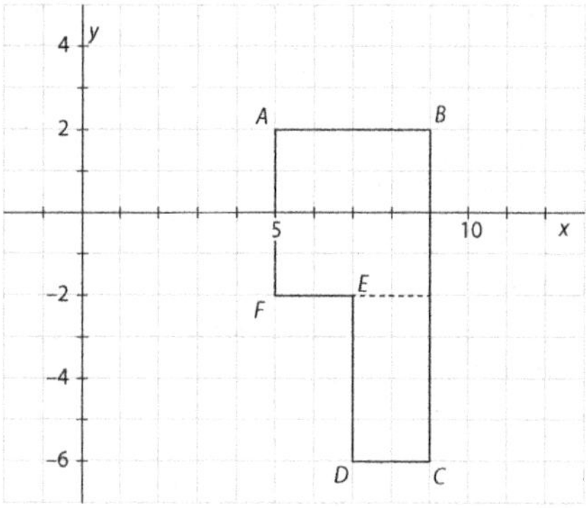

The figure's area, A, equals the sum of the areas of the square and rectangle:

$$\begin{aligned} A &= (4 \text{ units})(4 \text{ units}) + (2 \text{ units})(4 \text{ units}) \\ &= 16 \text{ units}^2 + 8 \text{ units}^2 \\ &= 24 \text{ units}^2 \end{aligned}$$

2. Enclose the triangle in a rectangle.

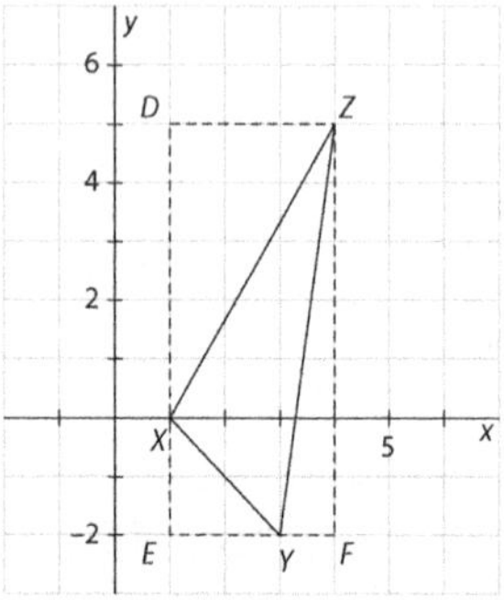

The area of triangle XYZ is the area of rectangle $DEFZ$ minus the areas of right triangles XDZ, XEY, and YFZ. In this figure, you can find the lengths of the sides of the figures by counting.

Rectangle $DEFZ$ has dimensions 7 units by 3 units. Triangle XDZ has base 3 units and height 5 units. Triangle XEY has base 2 units and height 2 units. And triangle YFZ has base 1 unit and height 7 units.

$$\begin{aligned} A &= (7\text{ units})(3\text{ units}) - \frac{1}{2}(3\text{ units})(5\text{ units}) - \frac{1}{2}(2\text{ units})(2\text{ units}) - \frac{1}{2}(1\text{ unit})(7\text{ units}) \\ &= 21\text{ units}^2 - 7.5\text{ units}^2 - 2\text{ units}^2 - 3.5\text{ units}^2 \\ &= 8\text{ units}^2 \end{aligned}$$

Triangle XYZ has area 8 units^2.

3. Cut the figure into a triangle and a square as shown below. Construct the altitude of the triangle and label its endpoint and coordinates.

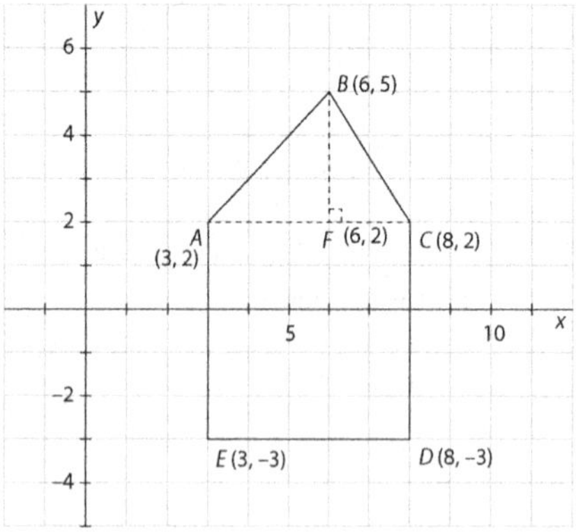

The area of the figure is the sum of the area of triangle ABC and the area of square $AEDC$. In this figure, find dimensions by counting. Triangle ABC has base 5 units and height 3 units. Square $AEDC$ has dimensions 5 by 5 units.

$$\begin{aligned} A &= \frac{1}{2}(5 \text{ units})(3 \text{ units}) + (5 \text{ units})(5 \text{ units}) \\ &= 7.5 \text{ units}^2 + 25 \text{ units}^2 \\ &= 32.5 \text{ units}^2 \end{aligned}$$

4. The distance Tyson runs is the perimeter of the path. Count the grid lengths. Going clockwise, from the start to the first corner is 4 grid lengths, to the next corner is 3, then 4, then 2, then 5, then 5, then 2, then 5, then 3, then 5, then 2. The total is 40 grid lengths. Multiply by 10 meters to obtain (40)(10 meters) = 400 meters. Tyson runs 400 meters when he runs once around the path.

Finding the Volume of Right Rectangular Prisms

(CCSS.Math.Content.6.G.A.2)

A **right rectangular prism** is a three-dimensional solid figure that has the shape of a box. It is composed of six **faces,** all of which are rectangles. The top and bottom rectangles are the **bases** and the other four rectangles are the **lateral faces.** The line segments where two adjacent faces intersect are the **edges.**

The **volume** of a right rectangular prism is the amount of space inside it. It has three dimensions: length, width, and height. When you use the dimensions to find its volume, the volume units are cubic units, such as cubic inches (in^3), cubic feet (ft^3), cubic miles (mi^3), cubic meters (m^3), cubic kilometers (km^3), cubic centimeters (cm^3), and cubic millimeters (mm^3).

A right rectangular prism has dimensions $2\frac{1}{4}$ inches by 2 inches by 4 inches. What is its volume?

There are at least two ways to find the volume of the prism.

Method 1: First, determine how many $\frac{1}{4}$- inch cubes can be packed inside the prism. Next, multiply this number by $\frac{1}{4}$ in $\times \frac{1}{4}$ in $\times \frac{1}{4}$ in $= \frac{1}{64}$ in^3, the volume of one $\frac{1}{4}$- inch cube.

Step 1. Layer the bottom of the prism with $\frac{1}{4}$- inch cubes. Nine will fit across the length (because $2\frac{1}{4}$ is $\frac{9}{4} = 9$ one-fourths). Eight will fit across the width (because 2 is $\frac{8}{4} = 8$ one-fourths). So, it takes 9 × 8 = 72 one-fourth-inch cubes to completely cover the bottom of the box. Each cube is $\frac{1}{4}$ inch in height. If you pack the prism with layers of 72 one-fourth-inch cubes, it will take 16 layers to pack it completely (because 4 is $\frac{16}{4} = 16$ one-fourths). It will take 72 × 16 = 1,152 one-fourth-inch cubes to pack the prism.

Step 2. The prism's volume, V, is

$$1{,}152 \times \frac{1}{64}\text{ in}^3 = \frac{1{,}152}{1} \times \frac{1}{64}\text{ in}^3 = \frac{1{,}152}{64}\text{ in}^3 = 18\text{ in}^3$$

Method 2: Use the formula $V = lwh$, where V is the volume of the rectangular prism, l is its length, w is its width, and h is its height.

$$V = lwh = \left(2\frac{1}{4}\text{ in}\right)(2\text{ in})(4\text{ in}) = \left(\frac{9}{\cancel{4}_1}\text{ in}\right)\left(\frac{2}{1}\text{ in}\right)\left(\frac{\cancel{4}^1}{1}\text{ in}\right) = 18\text{ in}^3$$

Try These

1. A right rectangular prism has dimensions 4 feet by $1\frac{1}{2}$ feet by $2\frac{1}{2}$ feet.
   (a) Use Method 1 to determine the volume of the prism.
   (b) Use Method 2 to determine the volume of the prism.
2. Determine the volume of the rectangular prism shown.

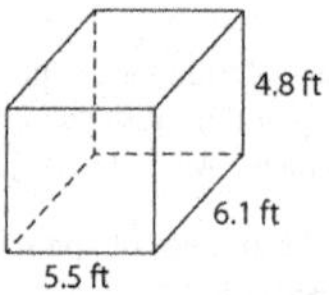

3. The owners of a house have added on a large room for entertaining family and friends. The room has dimensions 40 feet by 30 feet by 10 feet. Because of its size, the room will need its own air conditioning unit. The owners want to know what volume of air the air conditioner will have to cool. This information will determine the size of air conditioner they will purchase. Find the volume of the entertainment room.

Solutions

1. **(a)** *Step 1*. Layer the bottom of the prism with $\frac{1}{2}$- foot cubes. Eight will fit across the length (because 4 is $\frac{8}{2} = 8$ halves). Three will fit across the width (because $1\frac{1}{2}$ is $\frac{3}{2} = 3$ halves). So, it takes $8 \times 3 = 24$ half-foot cubes to completely cover the bottom of the box. Each cube is $\frac{1}{2}$ foot in height. If you pack the prism with layers of 24 half-foot cubes, it will take 5 layers to pack it completely (because $2\frac{1}{2}$ is $\frac{5}{2} = 5$ halves). It will take $5 \times 24 = 120$ half-foot cubes to pack the prism.

Step 2. Each half-foot cube has volume $\left(\frac{1}{2}\text{ ft}\right)\left(\frac{1}{2}\text{ ft}\right)\left(\frac{1}{2}\text{ ft}\right)=\frac{1}{8}\text{ ft}^3$. So, the prism's volume, V, is

$$120\times\frac{1}{8}\text{ ft}^3=\frac{120}{1}\times\frac{1}{8}\text{ ft}^3=\frac{120}{8}\text{ ft}^3=15\text{ ft}^3.$$

(b) $V=(4\text{ feet})\left(1\frac{1}{2}\text{ feet}\right)\left(2\frac{1}{2}\text{ feet}\right)=\left(\frac{\cancel{4}^{\,2}}{1}\text{ ft}\right)\left(\frac{3}{\cancel{2}_{\,1}}\text{ ft}\right)\left(\frac{5}{2}\text{ ft}\right)=\frac{30}{2}\text{ ft}^2=15\text{ ft}^2$

2. $V=(5.5\text{ ft})(6.1\text{ ft})(4.8\text{ ft})=161.04\text{ ft}^3$
3. The volume of the entertainment room is $V=(40\text{ ft})(30\text{ ft})(10\text{ ft})=12{,}000\text{ ft}^3$.

Using Nets to Find the Surface Area of Three-Dimensional Figures

(CCSS.Math.Content.6.G.A.4)

A **net** is a two-dimensional shape that can be folded to make a three-dimensional solid figure in which each face is a flat surface. Here are six three-dimensional solids and a corresponding net for each. ***Tip:*** Nets are not unique. A solid can have more than one net configuration.

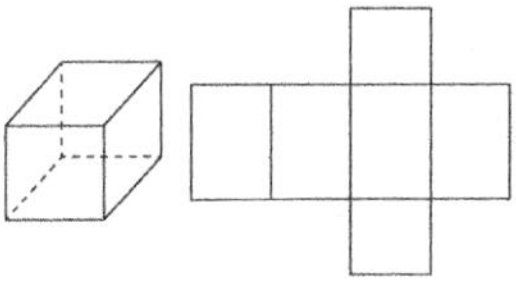

Right rectangular prism

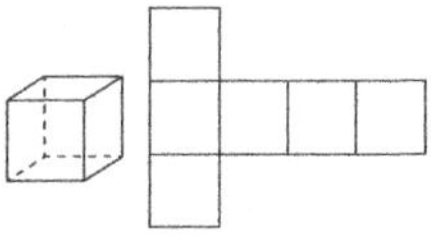

Cube

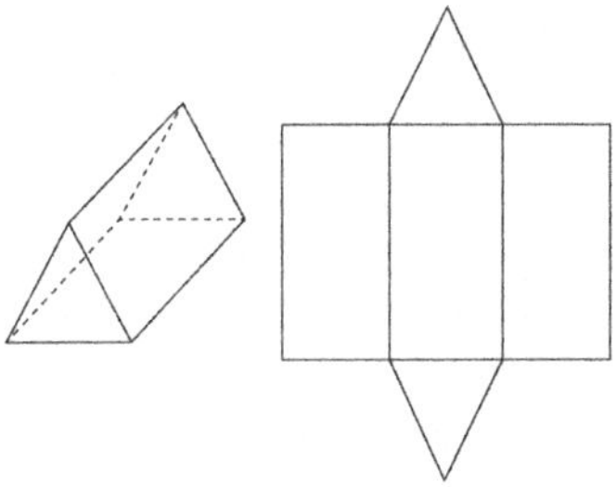

Triangular prism

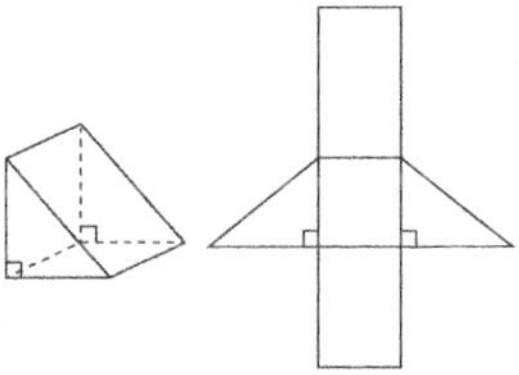

Right triangular prism

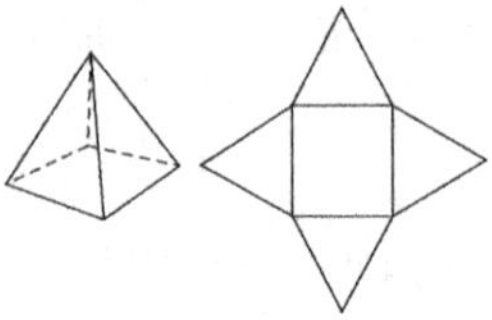

Square pyramid

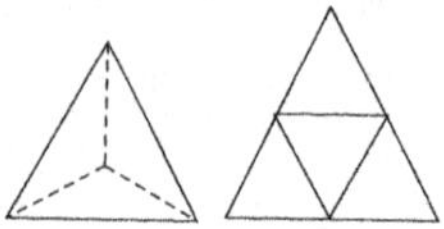

Triangular pyramid

Nets are helpful when you want to find the surface area of a solid figure. The **surface area** of a solid figure is the sum of the areas of its faces. Here are some examples.

Find the surface area of a 4-inch cube.

Sketch and label a net of the cube.

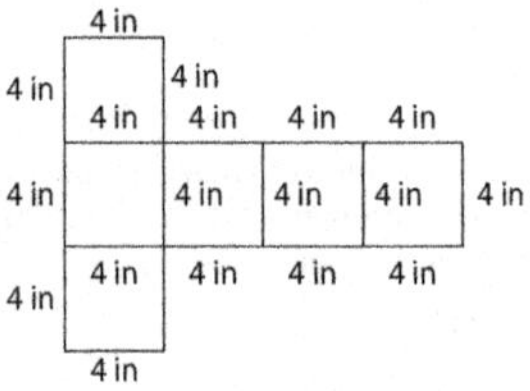

The surface area of the cube is the sum of the areas of its six square faces. Each face is a 4-inch by 4-inch square.

$$S.A. = 6(4\text{ in})(4\text{ in}) = 6(16\text{ in}^2) = 96\text{ in}^2$$

The grid shows the net for a square pyramid. Find the surface area of the pyramid. Assume each grid box represents a 1-inch by 1-inch square.

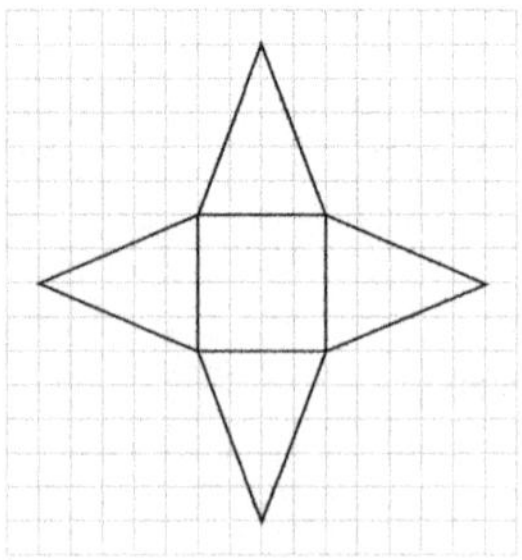

The surface area of the pyramid is the sum of the area of its square base and the areas of its four triangular faces. The square measures 4 inches by 4 inches. Each triangle has base 4 inches and height 5 inches.

$$S.A. = (4\text{ in})(4\text{ in}) + 4\left(\frac{1}{2}\right)(4\text{ in})(5\text{ in}) = 16\text{ in}^2 + 40\text{ in}^2 = 56\text{ in}^2$$

☞ Try These

1. Find the surface area of the right rectangular prism shown.

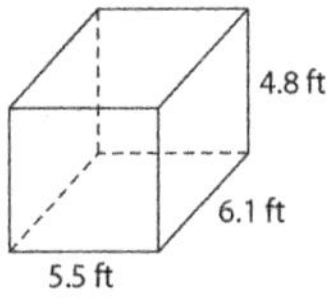

2. The figure shown is a net for a right triangular prism. Find the surface area of the prism.

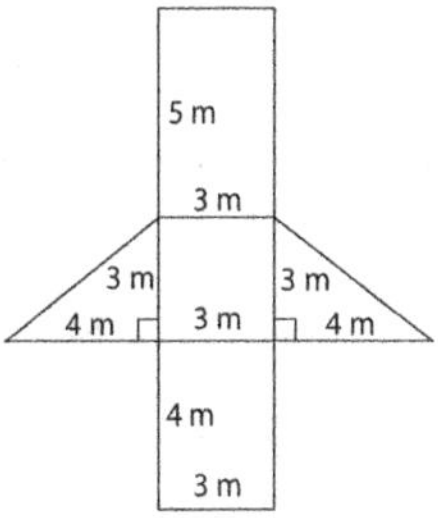

3. The grid shows the net for a triangular prism. Find the surface area of the prism. Assume each grid box represents a 1-centimeter by 1-centimeter square.

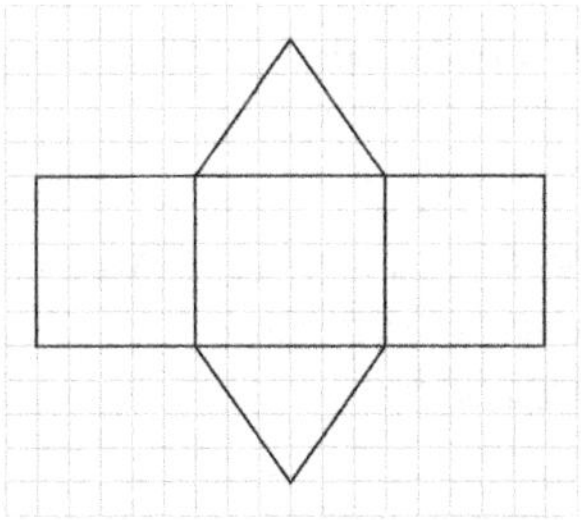

4. Maruja has a gift for a friend. She puts the gift in a plain box that has the form of a 5-inch cube. Then she wraps the box in colorful wrapping paper. How many square inches of wrapping paper will exactly cover the box?

5. The sum of the areas of the four triangular lateral faces of the right rectangular pyramid shown is approximately 248 cm^2. Find the pyramid's approximate surface area.

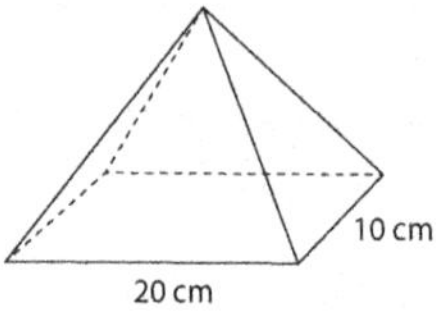

Solutions

1. Sketch and label the prism's net.

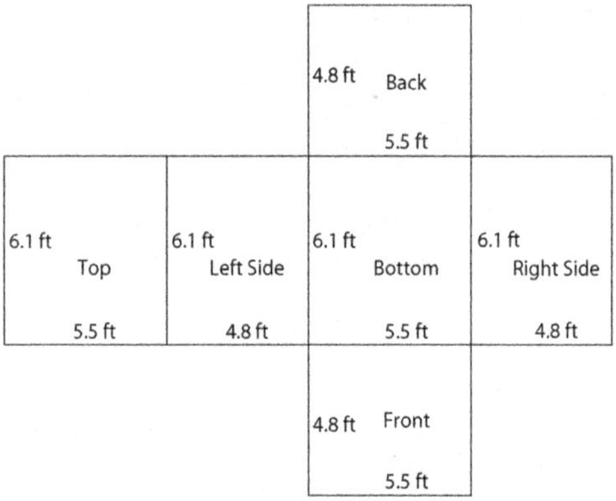

The prism's surface area is the sum of the areas of its six faces. Its top and bottom are 6.1-feet by 5.5-feet rectangles. Its front and back are 5.5-feet by 4.8-feet rectangles. And its sides are 6.1-feet by 4.8-feet rectangles.

$$\begin{aligned} S.A. &= 2(6.1\text{ ft})(5.5\text{ ft}) + 2(5.5\text{ ft})(4.8\text{ ft}) + 2(6.1\text{ ft})(4.8\text{ ft}) \\ &= 67.1\text{ ft}^2 + 52.8\text{ ft}^2 + 58.56\text{ ft}^2 \\ &= 178.46\text{ ft}^2 \end{aligned}$$

2. The prism's surface area is the sum of the areas of its five faces.

$$S.A. = (3\text{ m})(5\text{ m}) + (3\text{ m})(3\text{ m}) + (3\text{ m})(4\text{ m}) + 2\left(\frac{1}{2}\right)(3\text{ m})(4\text{ m})$$
$$= 15\text{ m}^2 + 9\text{ m}^2 + 12\text{ m}^2 + 12\text{ m}^2$$
$$= 48\text{ m}^2$$

3. The triangular prism's surface area is the sum of the areas of its three rectangular faces and its two congruent triangular faces. The center rectangle has dimensions 6 centimeters by 5 centimeters. The two outer rectangles are 5-centimeter squares. Each triangle has base 6 centimeters and height 4 centimeters.

$$S.A. = (6\text{ cm})(5\text{ cm}) + 2(5\text{ cm})(5\text{ cm}) + 2\left(\frac{1}{2}\right)(6\text{ cm})(4\text{ cm})$$
$$= 30\text{ cm}^2 + 50\text{ cm}^2 + 24\text{ cm}^2$$
$$= 104\text{ cm}^2$$

4. The cube has six faces. Each face is a 5-inch square.

$$S.A. = 6(5\text{ in})(5\text{ in}) = 150\text{ in}^2$$

To exactly cover the box will take 150 square inches of wrapping paper.

5. The surface area is the sum of the area of the pyramid's base and the areas of its four triangular lateral faces.

$$S.A. = (20\text{ cm})(10\text{ cm}) + 248\text{ cm}^2 = 200\text{ cm}^2 + 248\text{ cm}^2 = 448\text{ cm}^2$$

5. Statistics and Probability

In this chapter, you will learn about statistical questions and explore characteristics of distributions. You will create graphical displays of numerical data including dot plots, histograms, and box plots. You will use measures of center and variation to describe data. And you will summarize numerical data sets in relation to their contexts.

Recognizing Statistical Questions

(CCSS.Math.Content.6.SP.A.1)

A **statistical question** is one that anticipates the data collected to answer it will vary. It does not have a specific predetermined answer. For instance, "How tall are the sixth-grade students in your school?" is a statistical question. You expect the heights of sixth graders to vary from student to student. But "How tall is your best friend?" is *not* a statistical question. Your best friend has a specific height. There is no variability in the answer at the time of the question.

Tip: Statistical questions are answered by collecting data that vary.

☞ Try These

1. Tell whether or not the question posed is a statistical question. Justify your answer.
   (a) How many minutes do sixth graders in this class spend doing homework each day?
   (b) How many minutes do I spend doing homework each day?
2. Tell whether or not the question posed is a statistical question. Justify your answer.
   (a) How old is your social studies teacher?
   (b) How old are the sixth-grade teachers at your school?
3. Tell whether or not the question posed is a statistical question. Justify your answer.
   (a) How many eyes do humans have?
   (b) What is the most common eye color of humans?
4. Tell whether or not the question posed is a statistical question. Justify your answer.
   (a) What are the scores on the social studies test of students in this classroom?
   (b) What is my score on the social studies test?
5. Tell whether or not the question posed is a statistical question. Justify your answer.
   (a) Are sixth-grade girls taller than sixth-grade boys?
   (b) Is that sixth-grade girl taller than that sixth-grade boy?

Solutions

1. **(a)** This question is a statistical question. It is answered by collecting data that vary.
   (b) This question is not a statistical question. It has a specific predetermined answer. There is no variability in the answer at the time of the question.

2. **(a)** This question is not a statistical question. It has a specific predetermined answer. There is no variability in the answer at the time of the question.
   (b) This question is a statistical question. It is answered by collecting data that vary.

3. **(a)** This question is not a statistical question. It has a specific predetermined answer. There is no variability in the answer at the time of the question.
   (b) This question is a statistical question. It is answered by collecting data that vary.

4. **(a)** This question is a statistical question. It is answered by collecting data that vary.
   (b) This question is not a statistical question. It has a specific predetermined answer. There is no variability in the answer at the time of the question.

5. **(a)** This question is a statistical question. It is answered by collecting data that vary.
   (b) This question is not a statistical question. It has a specific predetermined answer. There is no variability in the answer at the time of the question.

Graphically Representing Data

(CCSS.Math.Content.6.SP.A.2, CCSS.Math.Content.6.SP.B.4, CCSS.Math.Content.6.SP.B.5.A, CCSS.Math.Content.6.SP.B.5.B)

The measurements or survey responses that are collected for the purpose of answering a statistical question are **data.** Consider the statistical question, "What are the scores on the social studies test of students in this classroom?" To collect data to answer this question, a sixth-grade social studies teacher gives and grades a 100-point test. The scores earned by the teacher's 20 students are listed below.

100, 95, 85, 30, 60, 75, 75, 80, 100, 65, 80, 95, 75, 70, 80, 75, 100, 75, 70, 70

These data are numerical data. **Numerical data** result from situations that involve taking measurements, such as lengths, heights, weights, or temperature, or situations in which objects or people are counted, such as the number of words on a page, the number of pets a child has, or the number of correct responses on a test.

Two useful graphical representations of numerical data are **dot plots** (or **line plots**) and **histograms.** These graphs provide visual representations of a data set's distribution to make analyzing the data easier. The **distribution** shows the shape and spread of the data.

By examining a dot plot or histogram, you can identify clusters, gaps, and outliers. A **cluster** is a group of data that are close together. A **gap** is an interval where no data are plotted. An **outlier** is a data value that is extremely high or extremely low in comparison to most of the other data values.

Understanding and Creating Dot Plots

A **dot plot** shows the frequency of data values on a number line. Dots (or other similar symbols) are placed above each value to indicate the number of times that particular value occurs in the data set. Here are the steps for creating a dot plot using the social studies test data.

100, 95, 85, 30, 60, 75, 75, 80, 100, 65, 80, 95, 75, 70, 80, 75, 100, 75, 70, 70

Step 1. Draw a horizontal axis. Mark equal intervals on the axis. Mark a scale for the axis that includes the least data value and the greatest data value. For the social studies test data, the least value is 30 and the greatest value is 100.

Step 2. Place one dot above the corresponding value on the horizontal axis for each value in the data set.

Step 3. Label the horizontal axis. For this example, use "Score" for the label.

Step 4. Title the graph. For this example, use "Social Studies Test Scores of 20 Sixth-Grade Students" as the title. ***Tip:*** Always report the number of data values in the title.

Here is the completed dot plot.

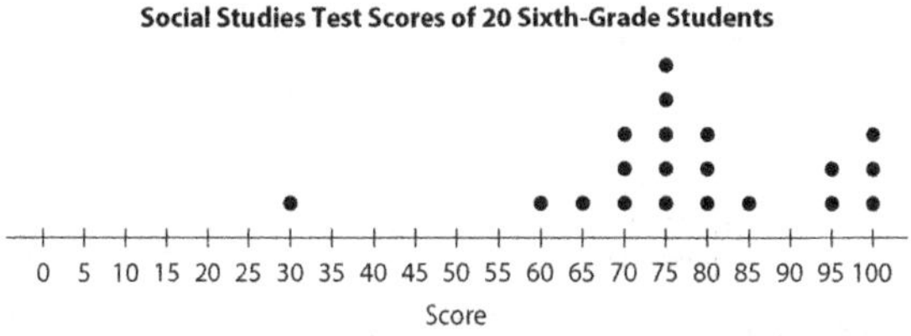

The dot plot provides visual information about the shape and spread of the data distribution. It shows a cluster of scores between 60 and 85 and a smaller cluster between 95 and 100. There is a large gap between 30 and 60 and a smaller one between 85 and 95. Because the gap between 30 and 60 is very large (indicating 30 is extremely low in comparison to the other data values), 30 is an outlier in the data set. The score that occurs most frequently is 75. The least score is 30 and the greatest score is 100.

An important advantage of dot plots is they visually show every data value. Also, they clearly display clusters, gaps, and outliers. You can easily determine the minimum (least) and the maximum (greatest) data values, the frequency of occurrence of values, and other useful information. However, dot plots are used mostly with small data sets (those with fewer than 50 data values).

☞ Try These

1. Fill in the blank.
   (a) A dot plot shows the __________ of data values on a number line.
   (b) An important advantage of dot plots is they visually show __________ data value.
   (c) Dot plots are used mostly with __________ (large, small) data sets.

2. A teacher collects data from the 20 students in her class to answer the statistical question, "How many minutes do students in this class spend getting dressed for school each day?" Each student estimated his or her own time to the nearest minute. Below are the data.

5, 5, 5, 7, 8, 8, 10, 10, 10, 10, 10, 10, 10, 15, 15, 15, 15, 20, 20, 45

(a) Create a dot plot to display the data.
(b) Describe the data's distribution by discussing clusters, gaps, and outliers.

Solutions

1. **(a)** frequency
(b) every
(c) small

2. **(a)**

(b) The dot plot shows a cluster of scores between 5 and 10. There is a very large gap between 20 and 45, indicating 45 is an outlier in the data set.

Understanding and Creating Histograms

A **histogram** is a vertical bar graph in which the bars are placed side-by-side with no space in between (except when no data occur between bars). It summarizes data by displaying frequencies of the data within specified intervals of equal size. Here are the steps for creating a histogram using the social studies test data.

100, 95, 85, 30, 60, 75, 75, 80, 100, 65, 80, 95, 75, 70, 80, 75, 100, 75, 70, 70

Step 1. Create five to six intervals of equal width that cover from the lowest to the highest data value. In this example, the difference between the highest score of 100 and the lowest score of 30 is $100 - 30 = 70$. Using five intervals, the width of each interval is $70 \div 5 = 14$.

Step 2. Make a frequency table of the data.

Interval of Scores	Tally	Frequency
30 to less than 44	|	1
44 to less than 58	0	0
58 to less than 72	𝍸	5
72 to less than 86	𝍸 ||||	9
86 to 100	𝍸	5

Step 3. Draw a horizontal axis. Mark the equal intervals on the axis.

Step 4. Draw a vertical axis. Mark a scale on the vertical axis that includes the low and high frequency counts. Make sure the units are equally spaced.

Step 5. Construct a bar above each interval so that the height of the bar corresponds to the frequency of occurrence of the data values within that interval.

Step 6. Label the horizontal axis. For this example, use "Score" for the label.

Step 7. Label the vertical axis "Frequency."

Step 8. Title the graph. For this example, use "Social Studies Test Scores of 20 Sixth-Grade Students" as the title.

Here is the completed histogram.

Social Studies Test Scores of 20 Sixth-Grade Students

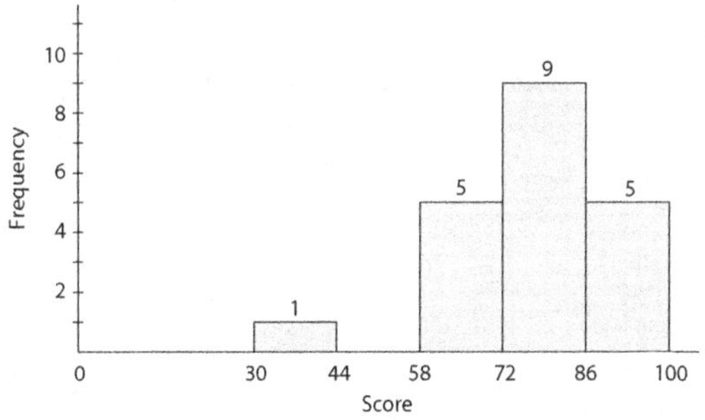

Tip: For the convenience of the reader, shade the bars in histograms and mark the frequency count at the top of each bar.

The histogram provides visual information about the shape and spread of the data distribution. It shows a cluster of scores between 58 and 100. The interval "72 to less than 86" has the greatest frequency. There is a fairly large gap between 44 and 58.

Because of the grouping of data into intervals, histograms do not provide the high level of specific information about data values that dot plots provide. The exact data values are not displayed in a histogram. For example, you know there is 1 data value in the interval "30 to less than 44," but without further information, you cannot determine its exact value. It could be any number from 30 up to 44. For the same reason, outliers and maximum and minimum values cannot be identified. Neither can the frequency of occurrence of particular values. Nevertheless, histograms are useful for displaying large sets of data, and therefore are used frequently in statistics.

In a **relative frequency histogram,** the scale is marked with relative frequencies instead of actual frequencies. You calculate the relative frequencies by dividing the frequency of occurrence in each interval by the sum of the frequencies. (***Tip:*** The sum of the frequencies is the same as the number of data values.) The relative frequency of an interval is the proportion of the data values that fall in the interval. The total of the relative frequencies corresponding to the intervals on the histogram should be 1.00 (but might instead be very close to 1.00 due to round-off error). Here are the relative frequencies for the social studies test data.

Interval of Scores	Tally	Frequency	Relative Frequency
30 to less than 44	|	1	$\frac{1}{20}=0.05$
44 to less than 58	0	0	0.00
58 to less than 72	~~||||~~	5	$\frac{5}{20}=0.25$
72 to less than 86	~~||||~~ ||||	9	$\frac{9}{20}=0.45$
86 to 100	~~||||~~	5	$\frac{5}{20}=0.25$

Here is the completed relative frequency histogram.

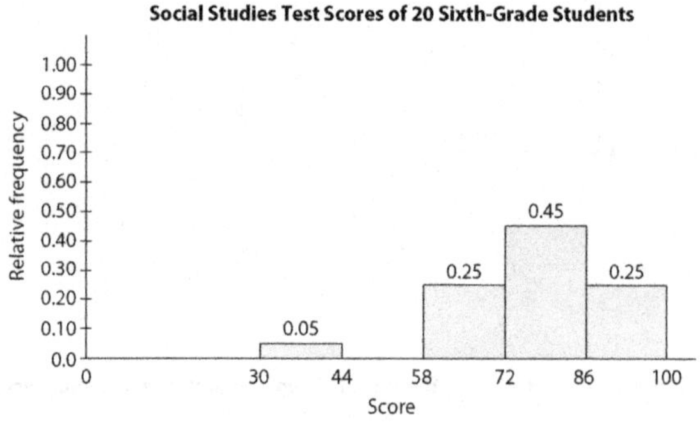

You also can show the relative frequencies as percentages as shown below.

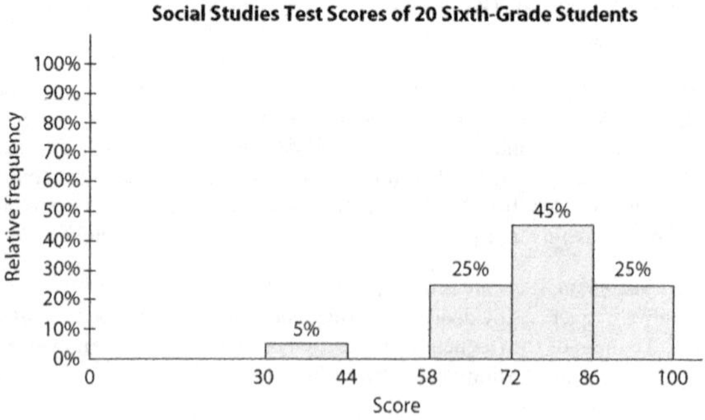

The only difference between the frequency histogram and the relative frequency histogram is the y-axis is scaled as proportions or percentages instead of actual frequency counts. The shape and spread of the two histograms are the same.

Tip: Relative frequency histograms are useful because proportions or percentages make it easier to consider portions of the data compared to the whole.

☞Try These

1. Fill in the blank.

   (a) A frequency histogram displays frequencies of the data within specified _________.
   (b) Because of the grouping of data, the exact data values _________ (are, are not) displayed in a histogram.
   (c) Histograms are used _________ (frequently, infrequently) in statistics.

2. The manager of an electronics store recorded the amounts spent by 200 customers who made purchases on the last day of a big sale. A partial frequency table of the data is shown below.

Interval of Amount Spent in Dollars	Tally	Frequency
0 to less than 100	卌 卌 卌 卌 卌 卌 卌 卌	
100 to less than 200	卌 卌 卌 卌 卌 卌 卌 卌 卌 III	
200 to less than 300	卌 卌 卌 卌 卌 卌 卌 卌 卌 卌	
300 to less than 400	卌 卌 卌 卌 卌 卌 IIII	
400 to less than 500	卌 卌 卌 I	
500 to less than 600	卌 卌 II	

   (a) Complete the frequency table for the manager's data.
   (b) Using the frequency table in part (a), create a histogram for the manager's data. Label the horizontal axis "Amount spent in dollars" and title the graph "The Amount Spent in Dollars by 200 Customers."
   (c) Describe the data's distribution by discussing clusters and gaps.

Solutions

1. **(a)** intervals
   (b) are not
   (c) frequently

2. **(a)**

Interval of Amount Spent in Dollars	**Tally**	**Frequency**
0 to less than 100	𝍸𝍸𝍸𝍸𝍸𝍸𝍸𝍸	40
100 to less than 200	𝍸𝍸𝍸𝍸𝍸𝍸𝍸𝍸𝍸𝍷𝍷𝍷	48
200 to less than 300	𝍸𝍸𝍸𝍸𝍸𝍸𝍸𝍸𝍸𝍸	50
300 to less than 400	𝍸𝍸𝍸𝍸𝍸𝍸𝍷𝍷𝍷𝍷	34
400 to less than 500	𝍸𝍸𝍸	16
500 to less than 600	𝍸𝍸𝍷𝍷	12

(b) Using the frequency table in part (a), create a histogram for the manager's data.

The Amount Spent in Dollars by 200 Customers

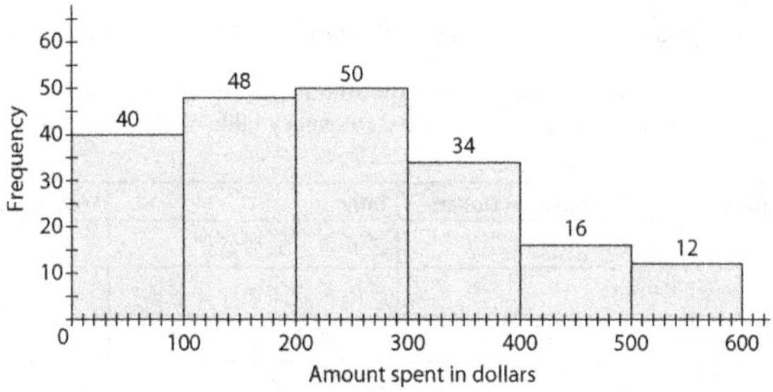

(c) The distribution appears to be clustered between 0 and 400. There are no gaps in the distribution.

Understanding and Determining Measures of Center

(CCSS.Math.Content.6.SP.A.2, CCSS.Math.Content.6.SP.A.3, CCSS.Math.Content.6.SP.B.4, CCSS.Math.Content.6.SP.B.5.B, CCSS.Math.Content.6.SP.B.5.C, CCSS.Math.Content.6.SP.B.5.D)

A **measure of center** is a numerical value that describes a data set by providing a "central" or "typical" value of the data set. It is a single number that summarizes all of the values in the data set.

The three most common measures of central tendency are the **mean, median,** and **mode.** Each of these measures represents a different way of describing a typical value of a set of data.

Tip: Measures of center should have the same units as those of the data values. If no units are specified, as in test scores, then the measure of center will not specify units.

Understanding and Determining the Mean of a Data Set

The **mean** of a data set is the arithmetic average of the values in the data set.

Determining the mean takes two steps. First, sum the values in the data set. Then, divide by how many numbers are in the set. Here is an example.

> Five friends earn money raking leaves on Saturdays. The amounts earned by the five friends last Saturday were \$50, \$80, \$50, \$30, and \$40. What is the mean amount earned by the five friends last Saturday?

Step 1. Sum the amounts.

$$\$50 + \$80 + \$50 + \$30 + \$40 = \$250$$

Step 2. Divide by 5.

$$\$250 \div 5 = \$50$$

The mean amount earned by the five friends last Saturday is \$50.

Think of the mean in terms of a "fair share." If the total amount of money were equally distributed among the five friends, each would get \$50.

Tip: In the real world, the mean is used extensively as a summary statistic for numerical data such as the mean weight, mean height, and mean score.

☞ Try These

1. Fill in the blank.

   (a) The mean of a data set is the arithmetic __________ of the values in the data set.
   (b) The mean is equivalent to a __________ share of the total of the data values.

2. Stephanie earned scores of 60, 95, 80, 75, 70, 65, 80, 100, 85, and 90 on ten daily quizzes. What is Stephanie's mean daily quiz score?

Solutions

1. **(a)** average
   (b) fair

2. *Step 1.* Sum the scores.

$$60 + 95 + 80 + 75 + 70 + 65 + 80 + 100 + 85 + 90 = 800$$

   Step 2. Divide by 10.

$$800 \div 10 = 80$$

   Stephanie's mean daily quiz score is 80.

Understanding and Determining the Median of a Data Set

The **median** is the middle value or the average of the two middle values in an ordered array of numerical data. Determining the median of a data set takes two steps. First, put the data values in order from least to greatest (or greatest to least). Then, find the middle value. If the number of data values is odd, the median is the middle value. If the number of data values is even, the median is the average of the two middle values. Here are some examples.

The ages in years of the nine children in Khamani's neighborhood are 13, 3, 10, 17, 15, 12, 12, 14, and 16. What is the median age of the children?

Step 1. List the ages from least to greatest.

3, 10, 12, 12, 13, 14, 15, 16, 17

Step 2. Find the middle value.

The number of data values is odd. The median is 13, the middle value.

The median age of the nine children is 13 years.

Tip: When you are finding a median, don't make the common mistake of neglecting to put the numbers in order first. In the preceding example, the middle number before the numbers are put in order is 15 (wrong answer).

A teacher asked six students to record the number of minutes each spent doing homework last week. Here are the data in minutes: 75, 250, 150, 450, 100, and 400. What is the median time spent doing homework last week?

Step 1. List the times from least to greatest.

75, 100, 150, 250, 400, 450

Step 2. Find the middle value.

The number of data values is even. The median is the average of 150 and 250, the two middle values. So, the median is $\frac{150+250}{2}=\frac{400}{2}=200$.

The median time the six children spent doing homework last week is 200 minutes.

Think of the median as the value that cuts the data in half. The number of data values below the median is the same as the number of data values above it.

☞ Try These

1. Fill in the blank.

   (a) If the number of data values is __________, the median is the middle value.
   (b) If the number of data values is even, the median is the __________ of the two middle values.
   (c) The median is the value that cuts the data in __________.

2. Shawn earned scores of 40, 95, 80, 75, 80, 65, 80, 100, 85, and 90 on ten daily quizzes. What is Shawn's median daily quiz score?

3. The weights in pounds of five sixth-grade boys are 80, 95, 84, 100, and 96. What is the median weight of the five boys?

Solutions

1. **(a)** odd
   (b) average
   (c) half

2. *Step 1.* List the scores from least to greatest.

   40, 65, 75, 80, 80, 80, 85, 90, 95, 100

   Step 2: Find the middle value.
   The number of data values is even. The median is the average of 80 and 80, the two middle values.
   Shawn's median daily quiz score is 80.

3. *Step 1.* List the weights from least to greatest.

   80, 84, 95, 96, 100

   Step 2: Find the middle value.
   The number of data values is odd. The median is 95, the middle value.
   The median weight of the five boys is 95 pounds.

Understanding and Determining the Mode of a Data Set

The **mode** is the number or numbers that occur with the greatest frequency in a set of data values. There can be one mode, more than one mode, or no mode. If two or more numbers occur with the same frequency that is greater than any of the other frequencies, then each will be a mode. When each number in the data set appears the same number of times, there is no mode.

Look at these examples.

There is one mode in the data set consisting of the values 50, 87, 50, 95, 78. The data value 50 occurs with the greatest frequency. Therefore, the mode is 50.

There are two modes in the data set consisting of the values 10, 10, 30, 36, 36, 100. The data values 10 and 36 both occur with the same frequency that is greater than any of the other frequencies. Therefore, the modes are 10 and 36.

There is no mode for the data set consisting of the values 40, 52, 145, 96, 60. Each data value in the data set appears the same number of times.

Here is a graphical example.

What score is the mode for the data shown in the dot plot below?

Social Studies Test Scores of 20 Sixth-Grade Students

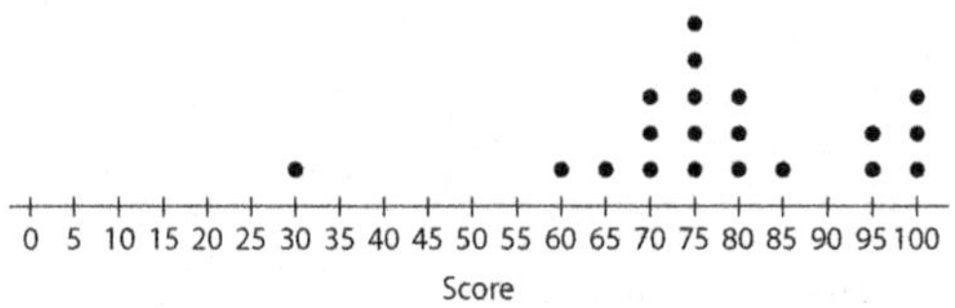

The data set displayed in the dot plot has one mode. The value 75 occurs five times, which is the greatest frequency. Therefore, 75 is the modal score of the 20 students.

Tip: The mode of a data set is the modal data value.

☞Try These

1. Fill in the blank.

   (a) The mode is the number or numbers that occur with the ________ frequency in a set of data values.

   (b) When each number in the data set appears the same number of times, there is ________ mode.

2. Stephanie earned scores of 60, 95, 80, 75, 70, 65, 80, 100, 85, and 90 on ten daily quizzes. What is Stephanie's modal daily quiz score?

3. The weights in pounds of five sixth-grade boys are 80, 95, 84, 100, and 96. What is the modal weight of the five boys?

Solutions

1. **(a)** greatest

   (b) no

2. The score of 80 occurs with the greatest frequency. Stephanie's modal daily quiz score is 80.

3. There is no mode because each weight occurs exactly one time.

Choosing Among the Mean, Median, and Mode

The mean, median, and mode are ways to describe a central or typical value of a data set. To know which of these measures of center you should use to describe a data set, consider these guidelines.

Mean

- The mean is preferred when the data distribution has a symmetric shape (or close to it).
- The actual data values are used in the computation of the mean. If any number is changed, the value of the mean will change. For example, the mean of the data set consisting of 50, 50, 87, 78, and 95 is 72. If the 95 in this set is changed to 100, the mean of the new data set is 73.
- A disadvantage of the mean is that it is influenced by outliers, especially in a small data set. It tends to be "pulled" toward an extreme value, much more so than does the median.
    - When a data set contains extremely high values that are not balanced by corresponding low values, the mean is misleadingly high. The mean of the data set consisting of 15, 15, 20, 25, and 25 is 20. If the 20 in this set is changed to 100, the mean of the new data set is 36. The value 36 does not represent the data set consisting of 15, 15, 100, 25, and 25 very well, since four of the data values are less than 30.
    - When a data set contains extremely low values that are not balanced by corresponding high values, the mean is misleadingly low. The mean of the data set consisting of 100, 100, 130, and 150 is 120. If the 150 in this set is changed to 10, the mean of the new data set is 85. The value 85 does not represent the data set consisting of 100, 100, 130, and 10 very well, since three of the data values are greater than or equal to 100.

Median

- The median is preferred when the data distribution is "lopsided" with unbalanced extreme values or outliers on one side. Such distributions are **skewed. Right-skewed** distributions have unbalanced extreme values on the right side. **Left-skewed** distributions have unbalanced extreme values on the left side.
- The median is not influenced by outliers. For instance, the median of the data set consisting of 10, 15, 20, 25, and 30 is 20. If the 30 in this set is changed to 100, the median of the new data set remains 20.
- A disadvantage of the median as an indicator of a central value is that it is based on relative size rather than on the actual numbers in the set. For instance, a student who has test scores of 44, 47, and 98 shows improved performance that would not be reflected if the median of 47, rather than the mean of 63, was reported as the representative score.

Mode

- The mode is the simplest measure of central tendency to calculate.
- If a data set has a mode, the mode (or modes) is one of the data values.
- The mode is the only appropriate measure of center for data that are strictly nonnumeric, such as data on eye color (blue, brown, green, or hazel). It makes no sense to determine a mean or median eye color. The eye color that occurred most frequently would be the modal eye color.
- For numeric data, the mean and median are preferred over the mode as measures of center.
- A disadvantage of the mode as an indicator of a central value is that it is based on relative frequency rather than on all the values in the set. For instance, a student who has test scores of 45, 45, and 99 shows improved performance that would not be reflected if the mode of 45, rather than the mean of 63, was reported as the representative score.

Try These

1. Fill in the blank.

   (a) The __________ is preferred when the data distribution has a symmetric shape.
   (b) The __________ is preferred when the data distribution is "lopsided" with unbalanced extreme values or outliers on one side.
   (c) The __________ is the simplest measure of central tendency to calculate.
   (d) The __________ is the only appropriate measure of center for data that are strictly nonnumeric.
   (e) A disadvantage of the mean is that it tends to be "pulled" toward a(n) __________ value, much more so than does the median.
   (f) The __________ is not influenced by outliers.

2. For the distributions shown, state which measure of center is preferred as a description of a typical value. Justify your answer.

   (a)

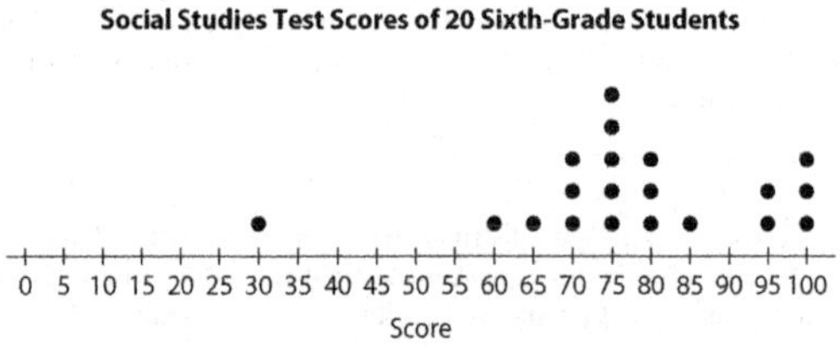

   (b)

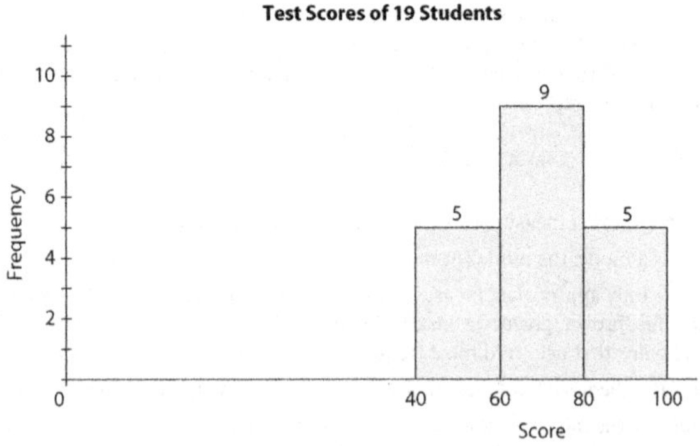

(c)

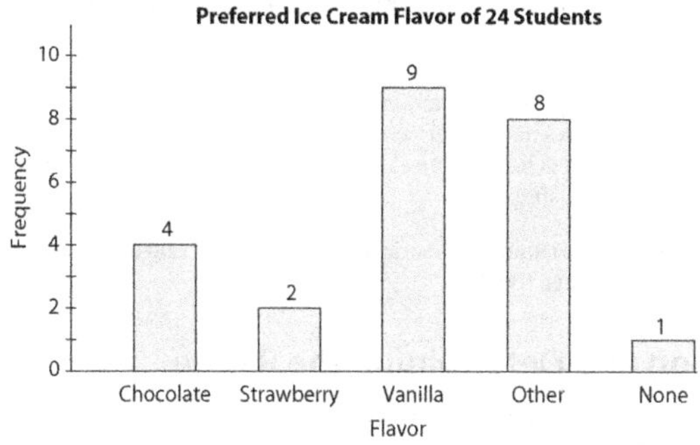

Solutions

1. **(a)** mean
   (b) median
   (c) mode
   (d) mode
   (e) extreme
   (f) median

2. **(a)** The median is the best choice as a measure of center for this distribution. The distribution is left-skewed. It has an outlier on the left side. The median is not influenced by extreme values.
   (b) This distribution has a symmetric shape, so the mean is the best choice as a measure of center for this distribution.
   (c) The data shown are strictly nonnumeric, so the mode is the best choice as a measure of center for this distribution.

Understanding and Determining Measures of Variation

(CCSS.Math.Content.6.SP.A.2, CCSS.Math.Content.6.SP.A.3, CCSS.Math.Content.6.SP.B.4, CCSS.Math.Content.6.SP.B.5.B, CCSS.Math.Content.6.SP.B.5.C, CCSS.Math.Content.6.SP.B.5.D)

Variation in data is of foremost interest to users of statistics. A **measure of variation** is a numerical value that describes the spread of a data set. Measures of center are important for describing data sets. However, their interpretation becomes more meaningful when the variation about the central value is known.

Tip: Measures of variation should have the same units as those of the data values. If no units are specified, then the measure of variation will not specify units.

For instance, one set of scores may be extremely consistent, with scores like 60, 62, 65, 68, 70, 70, 72, 75, 78, and 80; while another set of scores may be very erratic, with scores like 40, 40, 50, 55, 60, 80, 85, 90, 100, and 100. The scores in the first set cluster more closely together than do the scores in the second set; the scores in the second set are more spread out.

Three measures that quantify variation in numerical data sets are the **range,** the **mean absolute deviation (MAD),** and the **interquartile range (IQR).**

Understanding and Determining the Range

The **range** is the difference between the maximum (greatest) and the minimum (least) values in a data set. That is, range = maximum value – minimum value. The range describes how far apart the data are spread. Here is an example.

Lizbeth has run the 200-meter dash five times. Her times to the nearest tenth of a second are 29.3, 30.9, 22.8, 38.1, and 26.9. What is the range of Lizbeth's times?

$$\text{range} = \text{maximum value} - \text{minimum value} = 38.1 - 22.8 = 15.3$$

The range for Lizbeth's times is 15.3 seconds.

Try These

1. Fill in the blank.

   (a) The range is the __________ between the maximum and minimum values in a data set.
   (b) The range describes how far apart the data are __________.

2. Determine the range for the data shown in the dot plot.

Social Studies Test Scores of 20 Sixth-Grade Students

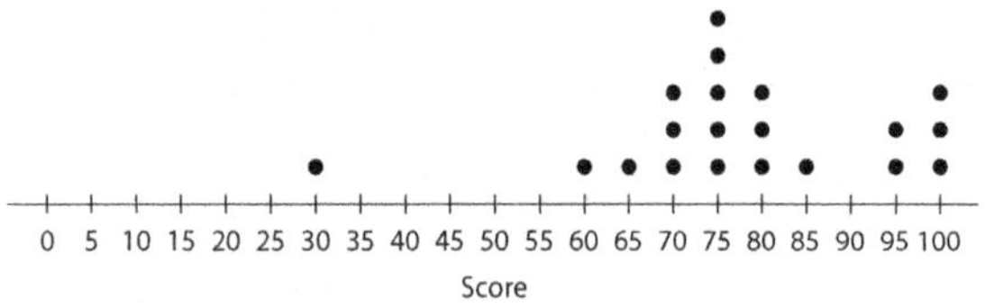

Solutions

1. **(a)** difference
   (b) spread
2. range = maximum value – minimum value = 100 – 30 = 70

Understanding and Determining the Mean Absolute Deviation (MAD)

The **MAD** is the average distance between each data value and the mean of the data values. It is a measure of spread that takes into account all the data values in the data set. If there is no variability in a data set, each data value equals the mean, so the MAD for the data set is zero. The more the data values vary from the mean, the greater the MAD.

Tip: The MAD is used for distributions in which the mean is the appropriate measure of center.

Determining the MAD takes three steps. First, determine the mean for the data values. Next, compute the sum of the distances of the data values from the mean. Then, divide the total distance from the mean of all values by the number of data values. Here is an example.

> Charlie reads during the week at night before bedtime. On Monday, Charlie read 70 minutes. On Tuesday, he read 30 minutes. On Wednesday, he read 60 minutes. On Thursday, he read 40 minutes. And on Friday, he read 50 minutes. What is the MAD for Charlie's reading times?

Step 1. Determine the mean. ***Tip:*** For convenience, list the times in order from least to greatest.

The mean is $(30 + 40 + 50 + 60 + 70) \div 5 = 250 \div 5 = 50$.

Step 2. Compute the sum of the distances of the data values from the mean.

Total distance $= (50 - 30) + (50 - 40) + (50 - 50) + (60 - 50) + (70 - 50) = 20 + 10 + 0 + 10 + 20 = 60$

Tip: A distance should be nonnegative. When you compute distances for the MAD, always subtract lesser values from greater values. This strategy will result in nonnegative distances for the MAD calculation.

Step 3. Divide by 5.

$$60 \div 5 = 12$$

The MAD for Charlie's reading times is 12 minutes. The MAD indicates the actual reading times for the five nights differ from the mean of 50 minutes by 12 minutes, on average.

When two data sets have the same mean, the MAD of the one whose data values are clustered closer to the mean is less than the MAD of the other data set. Here is an example.

> The following two data sets both have a mean of 50. Which data set has the greater MAD?

Set 1: 30, 40, 50, 60, 70

Set 2: 10, 10, 50, 90, 90

It is not necessary to calculate the actual MADs for Set 1 and Set 2. The two data sets have the same mean. The MAD of the one whose data values are clustered closer to the mean is less than the MAD of the other data set. The data values in Set 1 cluster more closely around the mean of 50 than do the data values in Set 2. So, The MAD of Set 1 is less than the MAD of Set 2. ***Tip:*** Even though the two data sets have the same mean, the data in Set 2 has more variability than does the data in Set 1.

Try These

1. Fill in the blank.

   (a) The MAD is the _________ distance between each data value and the mean of the data values.
   (b) The MAD is a measure of spread that takes into account _________ the data values in the data set.
   (c) The more the data values vary from the mean, the _________ (greater, lesser) the MAD.
   (d) The MAD is used for distributions in which the _________ is the appropriate measure of center.

2. What is the MAD for the data set consisting of 100, 100, 100, 100, 100, 100, 100, 100, 100, 100, 100, 100, 100, 100, 100, 100, 100, 100, 100, 100, 100, 100, and 100?

3. The weights in pounds of five sixth-grade girls are 70, 76, 84, 95, and 80. What is the MAD of the five girls' weights?

Solutions

1. **(a)** average
   (b) all
   (c) greater
   (d) mean

2. The data values are exactly the same. There is no variation in the data, so the MAD is zero.

3. *Step 1.* Determine the mean.
   The mean is (70 + 76 + 84 + 95 + 80) ÷ 5 = 405 ÷ 5 = 81.
   Step 2. Compute the sum of the distances of the data values from the mean.
   Total distance = (81 – 70) + (81 – 76) + (84 – 81) + (95 – 81) + (81 – 80) = 11 + 5 + 3 + 14 + 1 = 34
   Step 3. Divide by 5.

   $$34 \div 5 = 6.8$$

   The MAD for the girls' weights is 6.8 pounds. The MAD indicates the actual weights of the five girls differ from the mean of 81 pounds by 6.8 pounds, on average.

Understanding and Determining the Interquartile Range (IQR)

The **IQR** is the difference between the upper and lower quartiles of a data set. The **upper quartile** is the median of the upper half of the data set. The **lower quartile** is the median of the lower half of the data set. The IQR is the range of the middle 50 percent of the data. A small IQR indicates the middle half of the data cluster around the median. A large IQR indicates the middle half of the data are spread out away from the median.

Tip: The IQR is used for distributions in which the median is the appropriate measure of center.

Determining the IQR takes three steps. First, determine the data set's median. Next, determine the upper and lower quartiles. Then, compute the difference between the upper and lower quartiles. Here are some examples.

Determine the IQR for the mathematics test scores of 10 sixth-grade students.

95, 45, 60, 75, 100, 85, 90, 90, 95, 80

Step 1. Find the median of the 10 scores.

Put the data in order from least to greatest.

45, 60, 75, 80, 85, 90, 90, 95, 95, 100

The median is 87.5, the average of 85 and 90, the two middle values.

Step 2. Determine the upper and lower quartiles.

The upper quartile is 95, the median of 90, 90, 95, 95, 100.

The lower quartile is 75, the median of 45, 60, 75, 80, 85.

Step 3. Compute the difference between the upper and lower quartiles.

$$95 - 75 = 20$$

The IQR for the mathematics test scores of the 10 sixth-grade students is 20. The IQR indicates the bulk of the scores lie within a range of 20 points.

Determine the IQR for the ages in years of seven sixth-grade teachers.

35, 25, 42, 38, 27, 45, 52

Step 1. Find the median of the seven ages.

Put the data in order from least to greatest.

25, 27, 35, 38, 42, 45, 52

The median is 38, the middle value.

Step 2. Determine the upper and lower quartiles.

The upper quartile is 45, the median of 42, 45, 52.

The lower quartile is 27, the median of 25, 27, 35.

Tip: When the number of data values is odd, leave out the median to create the upper and lower halves of the data set.

Step 3. Compute the difference between the upper and lower quartiles.

$$45 - 27 = 18$$

The IQR for the ages of the seven sixth-grade teachers is 18 years. The IQR indicates the bulk of the ages are within a range of 18 years.

Try These

1. Fill in the blank.
   - **(a)** The IQR is the __________ between the upper and lower quartiles of a data set.
   - **(b)** The IQR is the range of the middle __________ percent of the data.
   - **(c)** The IQR is used for distributions in which the __________ is the appropriate measure of center.
2. The ages in years of the nine children in Khamani's neighborhood are 13, 3, 10, 17, 15, 12, 12, 14, and 16. What is the IQR for the ages of the children?
3. A teacher asked six students to record the number of minutes each spent doing homework last week. Here are the data in minutes: 75, 250, 150, 450, 100, and 400. What is the IQR for the times spent doing homework last week?

Solutions

1. **(a)** difference
   (b) 50
   (c) median

2. *Step 1.* Find the median of the nine ages.
   Put the data in order from least to greatest.

   3, 10, 12, 12, 13, 14, 15, 16, 17

   The median is 13, the middle value.

   Step 2. Determine the upper and lower quartiles.
   The upper quartile is 15.5, the median of 14, 15, 16, 17.
   The lower quartile is 11, the median of 3, 10, 12, 12.

   Step 3. Compute the difference between the upper and lower quartiles.

   $$15.5 - 11 = 4.5$$

   The IQR for the ages of the nine children is 4.5 years. The IQR indicates the bulk of the ages are within a range of 4.5 years.

3. *Step 1.* Find the median of the six times.

   Put the data in order from least to greatest.

   75, 100, 150, 250, 400, 450

   The median is 200, the average of 150 and 250, the two middle values.

Step 2. Determine the upper and lower quartiles.
The upper quartile is 400, the median of 250, 400, 450.
The lower quartile is 100, the median of 75, 100, 150.

Step 3. Compute the difference between the upper and lower quartiles.

$$400 - 100 = 300$$

The IQR for the times the six students spent doing homework is 300 minutes. The IQR indicates the bulk of the times are within a range of 300 minutes.

Determining the Five-Number Summary and Creating Box Plots

(CCSS.Math.Content.6.SP.A.4, CCSS.Math.Content.6.SP.B.A, CCSS.Math.Content.6.SP.B.5.C)

The **five-number summary** for a data set consists of five measures: the minimum data value (MIN), the lower quartile (Q1), the median (MED), the upper quartile (Q3), and the maximum data value (MAX). A **box plot** is a graphical representation of a five-number summary. It is a visual summary of a data set.

Tip: A box plot is also known as a box-and-whiskers plot.

Here is an example of creating a box plot from a five-number summary.

Create a box plot from the following five-number summary: MIN = 50, Q1 = 70, MED = 75, Q3 = 90, and MAX = 100.

Step 1. Draw a horizontal number line that fits the summary data.

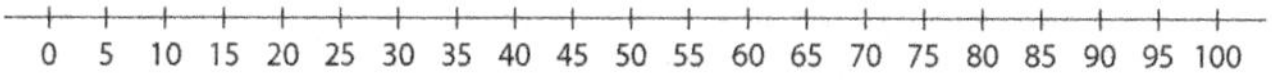

Step 2. Draw a rectangular box above the line (but not touching it) so that the left side of the box is directly above 70, the lower quartile (Q1), and the right side of the box is directly above 90, the upper quartile (Q3). Shade the box.

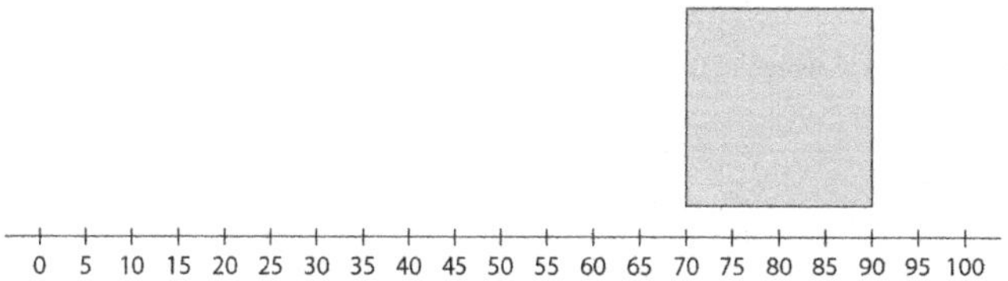

Step 3. Draw a vertical line within the box directly above 75, the median.

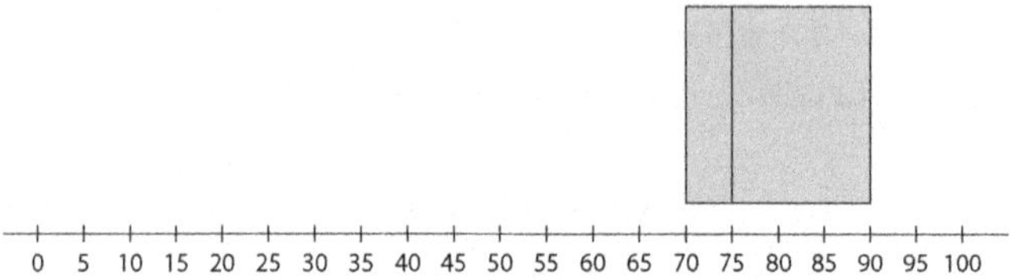

Step 4. Draw the "whiskers" for the box. Draw a horizontal line from the midpoint of the right side of the box extending right to a point directly above 100, the maximum value. Draw a horizontal line from the midpoint of the left side of the box extending left to a point directly above 50, the minimum value.

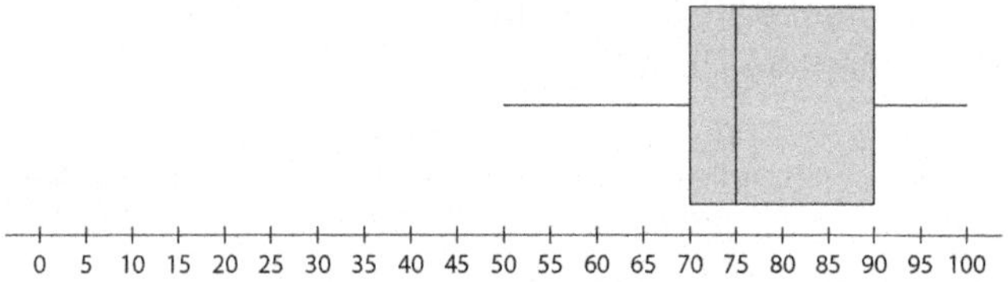

The box plot provides information about the location of the scores. Half of the scores are below 75 and half are above 75. Also the middle half of the scores falls between 70 and 90.

☞ Try These

1. Fill in the blank.

   (a) The five-number summary for a data set consists of the minimum data value, the lower quartile, the__________, the upper quartile, and the maximum data value.
   (b) A box plot is a visual __________ of a data set.
   (c) A box plot __________ (does, does not) show every data value.

2. The mathematics test scores of 10 sixth-grade students are shown below.

   95, 45, 60, 75, 100, 85, 90, 90, 95, 80

   Create a box plot to represent the sixth graders' mathematics test scores.

Solutions

1. **(a)** median
   (b) representation
   (c) does not

2. From an example in the previous section (see page 161), Q1 = 75, MED = 87.5, Q3 = 95. Inspection of the data gives MIN = 45 and MAX = 100.

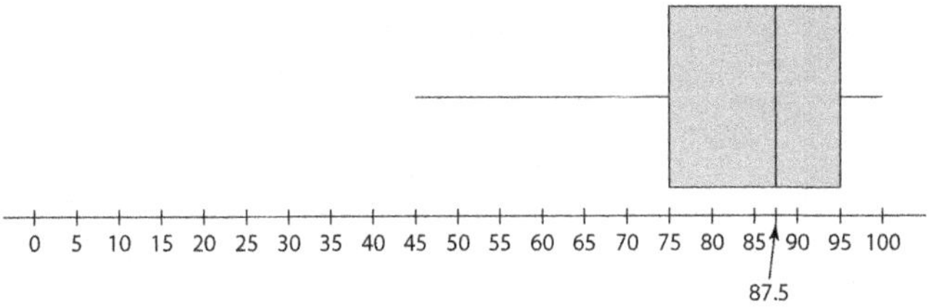

6. Practice Test 1

Directions: For questions 1–25, select the best answer choice.

1. A zoo has 13 peacocks and 25 parrots. What is the ratio of the number of peacocks to the total number of peacocks and parrots?

   A. 13 to 25
   B. 25 to 13
   C. 13 to 38
   D. 25 to 38

2. The ratio of the number of pencils Keenan has to the number of pencils Dominique has is 3 to 4. Keenan has 9 pencils. How many pencils does Dominique have?

   A. 4
   B. 6
   C. 8
   D. 12

3. The table below shows the number of calories for different amounts of white sugar.

Number of Teaspoons of Sugar	Number of Calories
2	32
5	80
7	?
10	160

   How many calories are in 7 teaspoons of sugar?
   A. 96
   B. 112
   C. 120
   D. 128

4. Shen paid \$69.30 for 14 frozen pepperoni pizzas. At this rate, how much would 18 of these pizzas cost?

   A. \$88.20
   B. \$89.10
   C. \$90.00
   D. \$91.80

5. In an auditorium of 300 students, 20% are sixth graders. How many of the students are sixth graders?

   A. 20
   B. 60
   C. 80
   D. 100

6. Lane is filling small bags with $\frac{3}{8}$ ounces of mixed spices. He has $5\frac{1}{4}$ ounces of mixed spices. How many bags can he fill?

A. 7
B. 14
C. 21
D. 28

7. Find the quotient of 58.5 and 1.125.

A. 0.52
B. 5.2
C. 52
D. 520

8. Garner wants to partition a rectangular 20-foot by 30-foot garden into smaller square plots. What is the length of the sides of the largest square plots into which Garner can partition the garden so that no land is left over?

A. 4 feet
B. 5 feet
C. 10 feet
D. 20 feet

9. Skye wants to mark the number $-\frac{4}{3}$ on the number line shown. Which statement is true about the location of $-\frac{4}{3}$ on the number line?

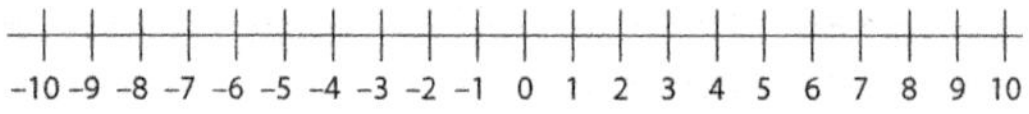

A. It is between –4 and –3.
B. It is between –3 and –2.
C. It is between –2 and –1.
D. It is between –1 and 0.

10. Which statement is true?

A. $-4.25 > -4.15$
B. $-3.92 < -3.5$
C. $-200 > -100$
D. $-2\frac{3}{4} > -2\frac{1}{4}$

11. Which list is in correct numerical order from least to greatest?

A. $-10\frac{3}{5}, -10.5, 0, \frac{3}{4}, \frac{4}{5}, 3.45, 3.8$

B. $-10.5, -10\frac{3}{5}, 0, \frac{3}{4}, \frac{4}{5}, 3.45, 3.8$

C. $-10\frac{3}{5}, -10.5, 0, \frac{4}{5}, \frac{3}{4}, 3.45, 3.8$

D. $-10\frac{3}{5}, -10.5, 0, \frac{3}{4}, \frac{4}{5}, 3.8, 3.45$

12. Order the absolute values of the numbers shown from least to greatest.

$$-5\frac{4}{5}, -40, -3.4, -3.215, -5\frac{1}{4}, -10$$

A. $-3.215, -3.4, -5\frac{1}{4}, -5\frac{4}{5}, -10, -40$

B. $-40, -10, -5\frac{4}{5}, -5\frac{1}{4}, -3.4, -3.215$

C. $3.4, 3.215, 5\frac{1}{4}, 5\frac{4}{5}, 10, 40$

D. $3.215, 3.4, 5\frac{1}{4}, 5\frac{4}{5}, 10, 40$

13. Name the quadrant in which the ordered pair (–4, 7) is located.

A. Quadrant I
B. Quadrant II
C. Quadrant III
D. Quadrant IV

14. Evaluate: $2 + 5^3 \times 3$

A. 47
B. 63
C. 377
D. 1,029

15. Evaluate the expression $5a + 1.5b$, when $a = 8.2$ and $b = 6$.

A. 37.3
B. 39
C. 42.5
D. 50

16. Which pair of expressions is equivalent?

A. $3(5x)$ and $8x$
B. $3(5x)$ and $15x$
C. $3x + 5x$ and $8x \cdot x$
D. $3x + 5x$ and $15x$

17. The original price of a wallet was decreased by $7.55. The price of the wallet after the decrease is $17.95. What was the original price of the wallet?

A. $9.40
B. $10.40
C. $24.50
D. $25.50

18. The length of a rectangular flower garden is 4.5 feet. The garden's area is 36 feet. What is the garden's width?

A. 4 feet
B. 6 feet
C. 8 feet
D. 9 feet

19. The table shows c, the number of calories, in n raw eggs. Write an equation that models the relationship between c and n.

Number of Eggs (*n*)	**Number of Calories (*c*)**
6	432
12	864
18	1,296

A. $c = \frac{1}{72}n$

B. $c = n + 72$
C. $n = 72c$
D. $c = 72n$

20. Determine the area of the given figure.

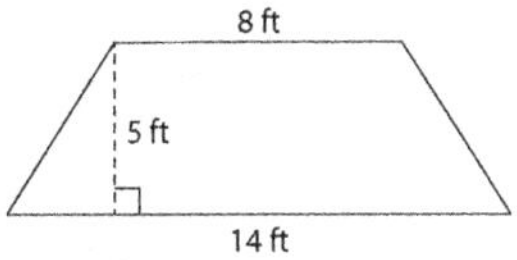

A. 40 ft^2
B. 55 ft^2
C. 70 ft^2
D. 112 ft^2

21. Determine the volume of the rectangular prism.

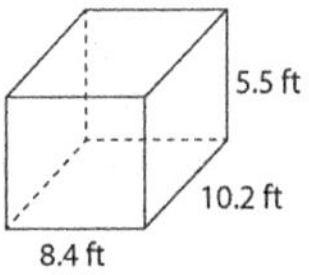

A. 48.2 ft^3
B. 72.3 ft^3
C. 235.62 ft^3
D. 471.24 ft^3

22. The figure shown is a net for a right triangular prism. Find the surface area of the prism.

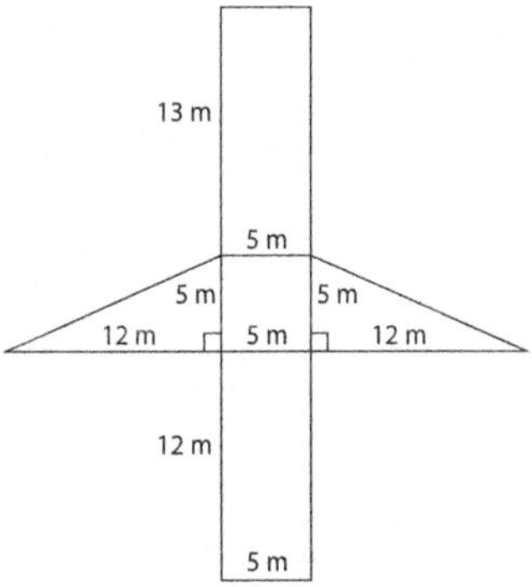

A. 180 m^2
B. 185 m^2
C. 210 m^2
D. 270 m^2

23. Which question is NOT a statistical question?

    A. For this school year, what is the salary of the teachers at this school?
    B. For this school year, what is my mathematics teacher's age?
    C. For this school year, what is the weight of boys in my class?
    D. For this school year, what is the height of girls in my class?

24. Sia earned scores of 65, 100, 95, 60, 75, 80, 70, 100, 85, and 90 on ten daily quizzes. What is Sia's median daily quiz score?

    A. 77.5
    B. 83.5
    C. 82.5
    D. 87.5

25. A fast-food restaurant gives a coupon for a free drink to customers who wait in line more than 15 minutes. The line plot shows the number of minutes 20 customers waited in line one morning at the restaurant.

Minutes 20 Customers Waited

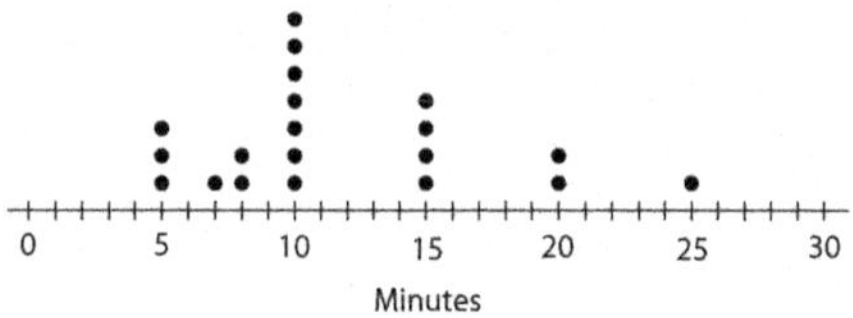

How many customers received a coupon for the free drink?

A. 2
B. 3
C. 4
D. 7

Directions: For questions 26–35, enter your answer in the answer box below the question. Enter the exact answer unless you are told to round your answer.

26. Isatu runs 725 meters in 5 minutes. At this rate, how many minutes will it take Isatu to run 290 meters?

[ ] minutes

27. Find the perimeter of rectangle *ABCD*.

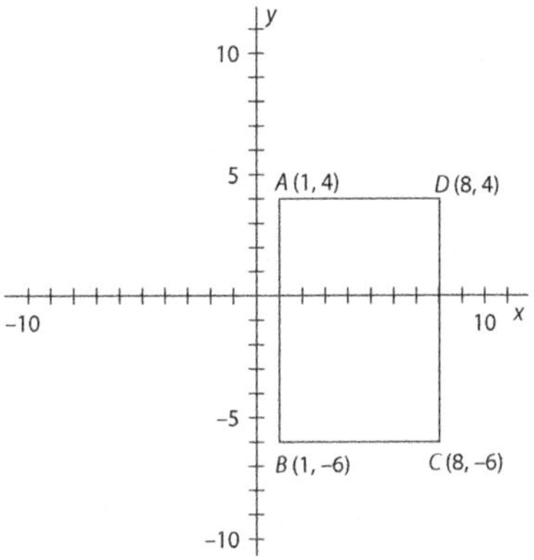

[ ] units

28. A customer gets a $7.99 discount off the original price of a shirt. The price of the shirt after the discount is $39.95. What was the shirt's original price?

$[ ]

29. A rectangular swimming pool has dimensions 30 by 20 feet. The pool is surrounded by a walkway that is 4 feet wide. Find the area of the walkway around the pool.

[ ] ft^2

30. Paula reads during the week at night before bedtime. On Monday, Paula read 60 minutes. On Tuesday, she read 40 minutes. On Wednesday, she read 50 minutes. On Thursday, she read 45 minutes. And on Friday, she read 35 minutes. What is the mean absolute deviation (MAD) for Paula's reading times?

[ ] minutes

31. The length of a rectangular piece of fabric is $\frac{2}{3}$ yard. The area of the fabric is $\frac{1}{2}$ square yard. What is the width of the fabric, in yards?

[ ] yard(s)

32. Compute: $5.4 - 3.875$

[ ]

33. The coordinates of point A are $(-8, 5)$. The coordinates of point B are $(4, 5)$. What is the distance between points A and B?

[ ] units

34. In the diagram of the quadrilateral below, the variables represent lengths, in inches. If $a = 4$ and $b = 12$, what is the perimeter of the quadrilateral, in inches?

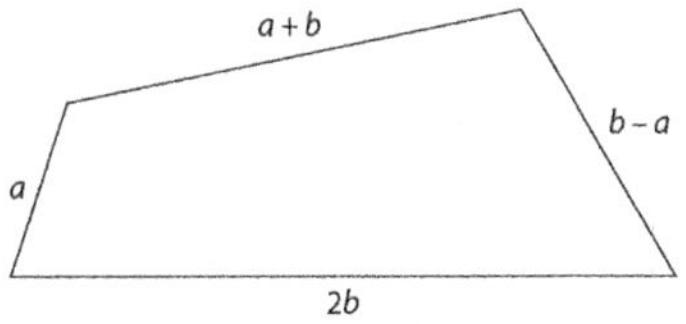

(figure not drawn to scale)

[ ] inches

35. In the diagram below, the sum of the measures of the three angles is 180°. What is the measure of $\angle x$ in degrees?

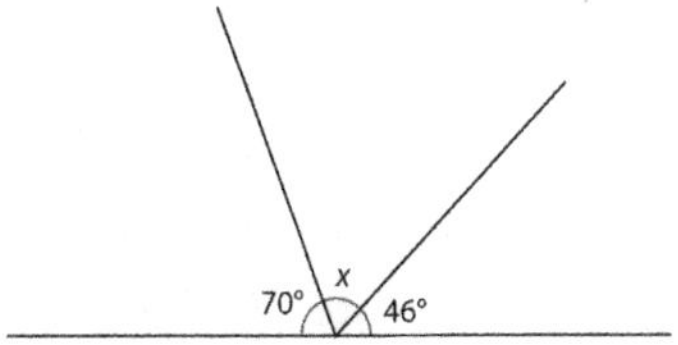

(figure not drawn to scale)

[ ]°

Answer Key

1. C
2. D
3. B
4. B
5. B
6. B
7. C
8. C
9. C
10. B
11. A
12. D
13. B
14. C
15. D
16. B
17. D
18. C
19. D
20. B
21. D
22. C
23. B
24. C
25. B
26. 2 minutes
27. 34 units
28. $47.94
29. 464 ft^2
30. 7.2 minutes
31. $\frac{3}{4}$ yard
32. 1.525
33. 12 units
34. 52 inches
35. 64°

Answer Explanations

1. **C.** Choice C is the correct response. Make a table to show the information in the question.

Number of Peacocks	Number of Parrots	Total
13	25	38

The total number of peacocks and parrots is 13 + 25 = 38. The ratio of the number of peacocks to the total is 13 to 38.

2. **D.** Choice D is the correct response.

Method 1: Use a tape diagram to solve the problem.

Keenan has 9 pencils. Dominque has 4 × 3 = 12 pencils as shown below.

Keenan | 3 | 3 | 3 | (3 parts)

Dominique | 3 | 3 | 3 | 3 | (4 parts)

Method 2: Create a ratio table. The table shows 9 pencils for Keenan corresponds to 12 pencils for Dominque.

Keenan's Number of Pencils	Dominique's Number of Pencils
3	4
6	8
9	12

Method 3: Write and solve an equation.
Let D = the number of pencils Dominque has and K = the number of pencils Keenan has.

Then $K = \frac{3}{4}D$ and $D = \frac{4}{3}K$. Substituting 9 for K in the second equation yields

$$D = \frac{4}{3}K = \frac{4}{3}(9) = \frac{4}{\cancel{3}_1}\left(\frac{\cancel{9}^3}{1}\right) = \frac{12}{1} = 12.$$

Dominique has 12 pencils.

3. **B.** Choice B is the correct response. Let S = the number of teaspoons of sugar. Let C = the number of calories. In the first row, the ratio of C to S is 32 to 2, which has value $\frac{32}{2} = \frac{16}{1}$. In the second row, the ratio of C to S is 80 to 5, which has value $\frac{80}{5} = \frac{16}{1}$. In the fourth row, the ratio of C to S is 160 to 10, which has value $\frac{160}{10} = \frac{16}{1}$. In the table, the number of calories is always 16 times the number of teaspoons. Using variable names, C is 16 times S. The equation is $C = 16S$. When S is 7 teaspoons of sugar, then $C = 16(7) = 112$.

The number of calories in 7 teaspoons of sugar is 112.

4. **B.** Choice B is the correct response. To solve the problem, use two steps. First, compute the price for one pizza. Next, multiply the price per pizza by 18.

   Step 1. Compute the price for one pizza.

$$\frac{\$69.30}{14 \text{ pizzas}} = \frac{\$69.30 \div 14}{(14 \div 14) \text{ pizzas}} = \frac{\$4.95}{1 \text{ pizza}} = \$4.95 \text{ per pizza}$$

   Step 2. Multiply the price per pizza by 18.

$$\frac{\$4.95}{1 \text{ pizza}} \times 18 \text{ pizzas} = \frac{\$4.95}{1 \cancel{\text{pizza}}} \times \frac{18 \cancel{\text{pizzas}}}{1} = \frac{\$89.10}{1} = \$89.10$$

   Eighteen pizzas will cost $89.10.

5. **B.** Choice B is the correct response. The percent is 20%. The whole is 300. The number of sixth graders is the part. The part is 20% of 300. The number of sixth graders (the part) is

$$20\% \text{ times } 300 = \frac{20}{100} \times 300 = \frac{1}{5} \times \frac{300}{1} = \frac{300}{5} = 60$$

   There are 60 sixth graders in the auditorium.

6. **B.** Choice B is the correct response. Determine how many $\frac{3}{8}$ ounces are in $5\frac{1}{4}$ ounces. To find the answer, divide $5\frac{1}{4}$ by $\frac{3}{8}$.

$$5\frac{1}{4} \div \frac{3}{8} = \frac{21}{4} \div \frac{3}{8} = \frac{{}^{7}\cancel{21}}{{}_{1}\cancel{4}} \times \frac{\cancel{8}^{2}}{\cancel{3}_{1}} = \frac{14}{1} = 14$$

   Lane can fill 14 bags.

7. **C.** Choice C is the correct response.

$$1.125\overline{)58.5} = 1.\underline{125})58.\underline{500} = 1125\overline{)58500.}$$

$$\begin{array}{r} 52. \\ 1125\overline{)58500.} \\ -5625 \\ \hline 2250 \\ -2250 \\ \hline 0 \end{array}$$

8. **C.** Choice C is the correct response. Make a sketch.

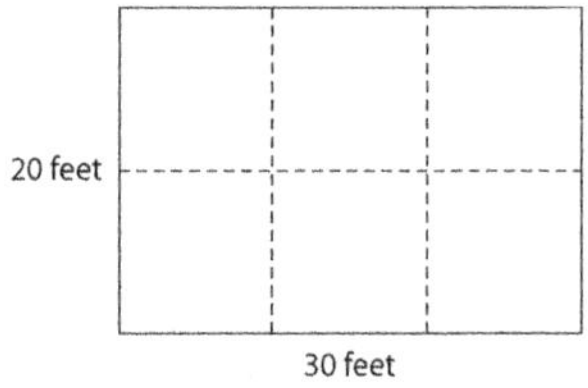

The length of the sides of the square plots into which Garner can partition the garden must be a factor of both 20 and 30. The largest such factor is the GCF of 20 and 30.

Factors of 20 are 1, 2, 4, 5, $\boxed{10}$, and 20; factors of 30 are 1, 2, 3, 5, 6, $\boxed{10}$, 15, and 30; the GCF is 10.

10 feet is the length of the sides of the largest square plots into which Garner can partition the garden so that no land is left over.

9. **C.** Choice C is the correct response. $-\frac{4}{3} = -1\frac{1}{3}$, so it lies between –2 and –1 on the number line.

10. **B.** Choice B is the correct response. Check each answer choice to determine which statement is true.

Checking A: Think: "*–4.25 is more negative than –4.15, so –4.25 < –4.15.*" Choice A is false.

Checking B: First, insert a zero after the last digit to the right of the decimal point in –3.5 to make the number of decimal places in the two numbers the same. Now compare –3.92 and –3.50. Think: "*–3.92 is more negative than –3.50, so –3.92 < –3.5.*" Choice B is true.

Because you know B is the correct response, you can go on to the next question. Here are the checks for C and D for your information.

Checking C: Think: "*–200 is more negative than –100, so –200 < –100.*" Choice C is false.

Checking D: Think: "$-2\frac{3}{4}$ *is more negative than* $-2\frac{1}{4}$, *so* $-2\frac{3}{4} < -2\frac{1}{4}$." Choice D is false.

11. **A.** Choice A is the correct response. Check each answer choice to determine which list is in correct numerical order.

Checking A: Rewrite each number (except 0) in the list as a two-place decimal.

$$-10\frac{3}{5},\ -10.5,\ 0,\ \frac{3}{4},\ \frac{4}{5},\ 3.45,\ 3.8 \rightarrow -10.60,\ -10.50,\ 0,\ 0.75,\ 0.80,\ 3.45,\ 3.80$$

The negative numbers to the left of zero are in the correct order because –10.60 is more negative than –10.50. The positive numbers to the right of zero are in the correct order. So,

$-10\frac{3}{5},\ -10.5,\ 0,\ \frac{3}{4},\ \frac{4}{5},\ 3.45,\ 3.8$, Choice A, is in correct numerical order.

Because you know A is the correct response, you can go on to the next question. Here are the checks for B, C, and D for your information.

Checking B: $-10.5,\ -10\frac{3}{5},\ 0,\ \frac{3}{4},\ \frac{4}{5},\ 3.45,\ 3.8$ is not in correct numerical order because –10.5 should be to the right of $-10\frac{3}{5}$.

Checking C: $-10\frac{3}{5},\ -10.5,\ 0,\ \frac{4}{5},\ \frac{3}{4},\ 3.45,\ 3.8$ is not in correct numerical order because $\frac{4}{5}$ should be to the right of $\frac{3}{4}$.

Checking D: $-10\frac{3}{5},\ -10.5,\ 0,\ \frac{3}{4},\ \frac{4}{5},\ 3.8,\ 3.45$ is not in correct numerical order because 3.8 should be to the right of 3.45.

12. **D.** Choice D is the correct response. The absolute values of the numbers are $5\frac{4}{5},\ 40,\ 3.4,\ 3.215,\ 5\frac{1}{4},\ 10$.

Eliminate choices A and B because absolute value is never negative. Choice C is incorrect because $3.215 < 3.4$.

13. **B.** Choice B is the correct response. The point is located 4 units to the left of the origin and 7 units above the x-axis. So, it lies in Quadrant II.

Tip: Remember, the quadrants are numbered *counterclockwise* starting with Quadrant I in the upper right.

14. **C.** Choice C is the correct response. Follow the order of operations.

$$\begin{aligned} 2+5^3\times 3 &= 2+125\times 3 \\ &= 2+375 \\ &= 377, \text{ Choice C} \end{aligned}$$

15. **D.** Choice D is the correct response.

$$\begin{aligned} 5a + 1.5b &= 5(a) + 1.5(b) \\ &= 5(8.2) + 1.5(6) \\ &= 41 + 9 \\ &= 50 \end{aligned}$$

16. **B.** Choice B is the correct response. Check the answer choices.

    Checking A: $3(5x)$ and $8x$ are not equivalent. If $x = 2$, $3(5x) = 3(5 \cdot 2) = 3(10) = 30$, but $8x = 8 \cdot 2 = 16$.

    Checking B: $3(5x)$ and $15x$ are equivalent. By the associative property of multiplication, $3(5x) = (3 \cdot 5)x = 15x$.

    Because you know B is the correct response, you can go on to the next question. Here are the checks for C and D for your information.

    Checking C: $3x + 5x$ and $8x \cdot x$ are not equivalent. If $x = 2$, $3x + 5x = 3 \cdot 2 + 5 \cdot 2 = 6 + 10 = 16$, but $8x \cdot x = 8 \cdot 2 \cdot 2 = 32$.

    Checking D: $3x + 5x$ and $15x$ are not equivalent. If $x = 2$, $3x + 5x = 3 \cdot 2 + 5 \cdot 2 = 6 + 10 = 16$, but $15x = 15 \cdot 2 = 30$.

17. **D.** Choice D is the correct response. Let c = the original price of the wallet in dollars. Write an equation that represents the facts given in the problem.

$$c - \$7.55 = \$17.95$$

    Solve the equation.

$$\begin{aligned} c - \$7.55 &= \$17.95 \\ c - \$7.55 + \$7.55 &= \$17.95 + \$7.55 \\ c &= \$25.50 \end{aligned}$$

    The wallet's original price was $25.50, Choice D.

18. **C.** Let w = the garden's width. The area, A, of a rectangle is $A = lw$, where l is the rectangle's length and w is its width. Write an equation that represents the facts given in the problem.

$$\begin{aligned} lw &= A \\ (4.5\text{ ft})(w) &= 36\text{ ft}^2 \end{aligned}$$

    Solve the equation. For convenience, omit the units while solving.

$$\begin{aligned} (4.5)(w) &= 36 \\ 4.5w &= 36 \\ \frac{4.5w}{4.5} &= \frac{36}{4.5} \\ \frac{{}^{1}\cancel{4.5}w}{{}_{1}\cancel{4.5}} &= 36 \div 4.5 \\ w &= 8 \end{aligned}$$

    The garden's width is 8 feet, Choice C.

19. **D.** Choice D is the correct response. Determine the ratio of c to n.

$\frac{432}{6}=\frac{72}{1}$, $\frac{864}{12}=\frac{72}{1}$, and $\frac{1{,}296}{18}=\frac{72}{1}$. So, c, the number of calories, is 72 times n, the number of eggs.

The equation is $c = 72n$, Choice D.

20. **B.** Choice B is the correct response. Cut the figure into two right triangles and a rectangle as shown here.

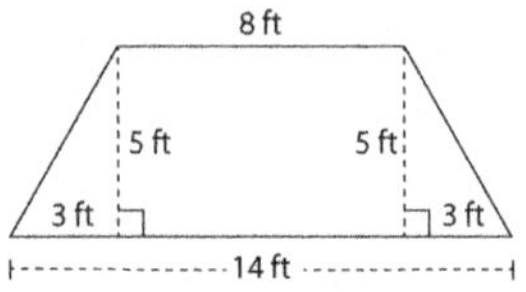

$$A=2\left(\frac{1}{2}\right)(3\text{ ft})(5\text{ ft})+(8\text{ ft})(5\text{ ft})=15\text{ ft}^2+40\text{ ft}^2=55\text{ ft}^2$$

21. **D.** Choice D is the correct response.

$$V = lwh = (8.4\text{ ft})(10.2\text{ ft})(5.5\text{ ft}) = 471.24\text{ ft}^3$$

22. **C.** Choice C is the correct response. The prism's surface area is the sum of the areas of its three rectangular sides and its two congruent right triangular bases.

$$\begin{aligned} S.A. &= (5\text{ m})(13\text{ m})+(5\text{ m})(5\text{ m})+(5\text{ m})(12\text{ m})+2\left(\frac{1}{2}\right)(12\text{ m})(5\text{ m}) \\ &= 65\text{ m}^2+25\text{ m}^2+60\text{ m}^2+60\text{ m}^2 \\ &= 210\text{ m}^2 \end{aligned}$$

23. **B.** Choice B is the correct response. The question in B is not a statistical question. It has a predetermined answer. There is no variability in the answer at the time of the question. The questions in A, C, and D are statistical questions. They are answered by collecting data that vary.

24. **C.** Choice C is the correct response.

Step 1. List the scores from least to greatest.

60, 65, 70, 75, 80, 85, 90, 95, 100, 100

Step 2. Find the middle value.

The number of data values is even.

The median is the average of 80 and 85, the two middle values. The median $=\frac{80+85}{2}=\frac{165}{2}=82.5$.

Sia's median daily quiz score is 82.5, Choice C.

25. **B.** Choice B is the correct response. The line plot shows 2 customers waited 20 minutes and 1 customer waited 25 minutes, for a total of 3 customers who waited more than 15 minutes.

So, 3 customers received a coupon for the free drink.

26. **2 minutes**

Step 1. Compute the rate for 1 minute.

$$\frac{725 \text{ meters}}{5 \text{ minutes}} = \frac{(725 \div 5) \text{ meters}}{(5 \div 5) \text{ minutes}} = \frac{145 \text{ meters}}{1 \text{ minute}}$$

Step 2. Multiply 290 meters by the equivalent rate of $\frac{1 \text{ minute}}{145 \text{ meters}}$.

$$\begin{aligned} 290 \text{ meters} \times \frac{1 \text{ minute}}{145 \text{ meters}} &= \frac{290 \cancel{\text{ meters}}}{1} \times \frac{1 \text{ minute}}{145 \cancel{\text{ meters}}} \\ &= \frac{290 \text{ minutes}}{145} \\ &= \frac{(290 \div 145) \text{ minutes}}{145 \div 145} \\ &= \frac{2 \text{ minutes}}{1} \\ &= 2 \text{ minutes} \end{aligned}$$

It will take Isatu 2 minutes to run 290 meters.

27. **34 units**

The perimeter of rectangle $ABCD = AB + BC + CD + AD$.

$$AB = CD = |4| + |-6| = 4 + 6 = 10$$
$$BC = AD = |8| - |1| = 8 - 1 = 7$$
$$\text{perimeter} = 10 + 10 + 7 + 7 = 34 \text{ units}$$

28. **\$47.94**

Let s = the shirt's original price in dollars. Write an equation that represents the facts given in the problem.

$$s - \$7.99 = \$39.95$$

Solve the equation.

$$\begin{aligned} s - \$7.99 &= \$39.95 \\ s - \$7.99 + \$7.99 &= \$39.95 + \$7.99 \\ s &= \$47.94 \end{aligned}$$

The shirt's original price was \$47.94.

29. **464 ft²**

Make a sketch.

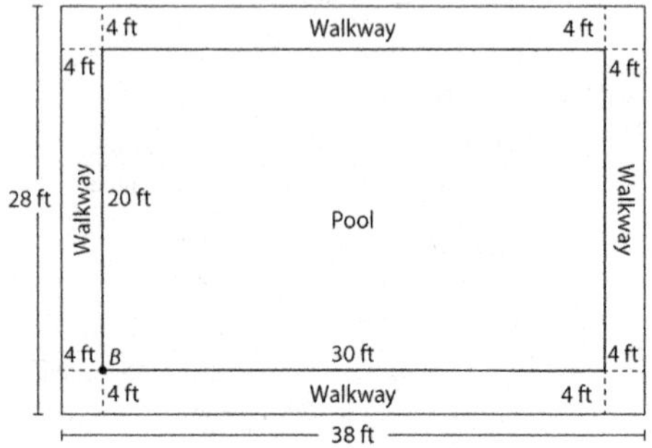

The area of the walkway is the difference between the area of the larger rectangle and the area of the smaller 30 ft-by-20 ft rectangle. The length of the larger rectangle is 30 ft + 2(4 ft) = 30 ft + 8 ft = 38 ft. Its width is 20 ft + 2(4 ft) = 20 ft + 8 ft = 28 ft. The area of the walkway is (38 ft)(28 ft) – (30 ft)(20 ft) = 1,064 ft^2 – 600 ft^2 = 464 ft^2.

30. **7.2 minutes**

The MAD is the average distance between each data value and the mean of the data values.

Step 1. Determine the mean reading time. ***Tip:*** For convenience, list the times in order from least to greatest.

$$\text{mean} = (35 + 40 + 45 + 50 + 60) \div 5 = 230 \div 5 = 46$$

Step 2. Compute the sum of the distances of the data values from the mean.

$$(46 - 35) + (46 - 40) + (46 - 45) + (50 - 46) + (60 - 46) = 11 + 6 + 1 + 4 + 14 = 36$$

Step 3. Divide by 5.

$$36 \div 5 = 7.2$$

The MAD for Paula's reading times is 7.2 minutes.

31. $\frac{3}{4}$ **yard**

The area of a rectangle is lw, where l is the rectangle's length and w is its width. Let w = the width of the fabric in yards.

Write an equation that represents the facts given in the problem.

$$lw = \frac{1}{2} \text{ square yard}$$
$$\left(\frac{2}{3} \text{ yard}\right)w = \frac{1}{2} \text{ square yard}$$

Solve the equation (omit the units for convenience).

$$\left(\frac{2}{3}\right)w = \frac{1}{2}$$
$$\left(\frac{3}{2}\right)\left(\frac{2}{3}\right)w = \left(\frac{3}{2}\right)\left(\frac{1}{2}\right)$$
$$\left(\frac{{}^{1}\not{3}}{{}_{1}\not{2}}\right)\left(\frac{\not{2}^{1}}{\not{3}_{1}}\right)w = \left(\frac{3}{2}\right)\left(\frac{1}{2}\right)$$
$$lw = \frac{3}{4}$$
$$w = \frac{3}{4}$$

The width of the fabric is $\frac{3}{4}$ yard.

32. **1.525**

$$\begin{array}{r} 5.400 \\ -3.875 \\ \hline 1.525 \end{array}$$

33. **12 units**

The points A and B have the same y-coordinate, so they lie on the same horizontal line. The x-coordinates have different signs. To find the distance between the two points, add the absolute values of the x-coordinates.

$$|-8| + |4| = 8 + 4 = 12$$

The distance is 12 units.

34. **52 inches**

The perimeter = $(a) + (2b) + (b - a) + (a + b) = a + 2b + b - a + a + b = a + 2b + 2b = a + 4b =$ $(4) + 4(12) = 4 + 48 = 52$.

The perimeter is 52 inches.

35. **64°**

Write an equation that represents the facts given in the problem.

$$x + 70° + 46° = 180°$$

Solve the equation (omit the units for convenience).

$$x + 70 + 46 = 180$$
$$x + 116 = 180$$
$$x + 116 - 116 = 180 - 116$$
$$x = 64$$

The measure of $\angle x$ is 64°.

7. Practice Test 2

Directions: For questions 1–25, select the best answer choice.

1. Write a ratio in simplest form for the ratio of 2 gallons to 10 quarts.

   A. 2:5
   B. 5:2
   C. 4:5
   D. 5:4

2. Which ratio is equivalent to 12 to 18?

   A. 18 to 24
   B. 12 to 30
   C. 8 to 12
   D. 4 to 9

3. A pancake recipe requires 3 cups of flour for 20 pancakes. How many cups of flour are needed to make 50 pancakes?

   A. 7 cups
   B. $7\frac{1}{2}$ cups
   C. 8 cups
   D. $8\frac{1}{2}$ cups

4. The ratio of the number of girls to the number of boys in a club is 8 to 5. The number of boys in the club is 15. How many girls are in the club?

   A. 24
   B. 20
   C. 16
   D. 8

5. A mixture of salt and water is 5% salt by weight. The amount of salt is 15 grams. How many grams does the mixture weigh?

   A. 300 grams
   B. 250 grams
   C. 200 grams
   D. 150 grams

6. How many $\frac{2}{3}$-cup servings are in $2\frac{2}{3}$ cups of rice?

   A. 3
   B. 4
   C. 6
   D. 8

7. Find the quotient of 2.5 and 0.005.

   A. 0.002
   B. 0.2
   C. 50
   D. 500

8. Brenton and Jaomi are walking around a circular 500-meter track. They both start at the same place and time and go in the same direction. Brenton walks at a steady pace and takes 4 minutes to go once around the track. Jaomi walks at a steady pace and takes 6 minutes to go once around. After how many minutes will they meet again at the starting point?

   A. 6
   B. 10
   C. 12
   D. 24

9. Alberto wants to mark the number –5.25 on the number line shown. Which statement is true about the location of –5.25 on the number line?

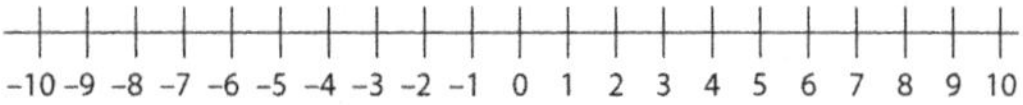

   A. It will lie to the left of –5.2.
   B. It will lie to the left of –5.3.
   C. It will lie to the left of –5.4.
   D. It will lie to the left of –5.5.

10. Which statement is true?

   A. $|-8.41| > |-14.15|$
   B. $|-10.62| > |9.5|$
   C. $|100| > |-200|$
   D. $\left|-5\frac{1}{4}\right| > \left|5\frac{1}{2}\right|$

11. Order the numbers shown from least to greatest.

$$4.1,\ 4.\overline{1},\ 4\frac{1}{11},\ 4,\ 4.11$$

A. $4.1,\ 4.\overline{1},\ 4\frac{1}{11},\ 4,\ 4.11$

B. $4,\ 4\frac{1}{11},\ 4.1,\ 4.\overline{1},\ 4.11$

C. $4,\ 4.1,\ 4\frac{1}{11},\ 4.11,\ 4.\overline{1}$

D. $4,\ 4\frac{1}{11},\ 4.1,\ 4.11,\ 4.\overline{1}$

12. Triangle *ABC* is shown below.

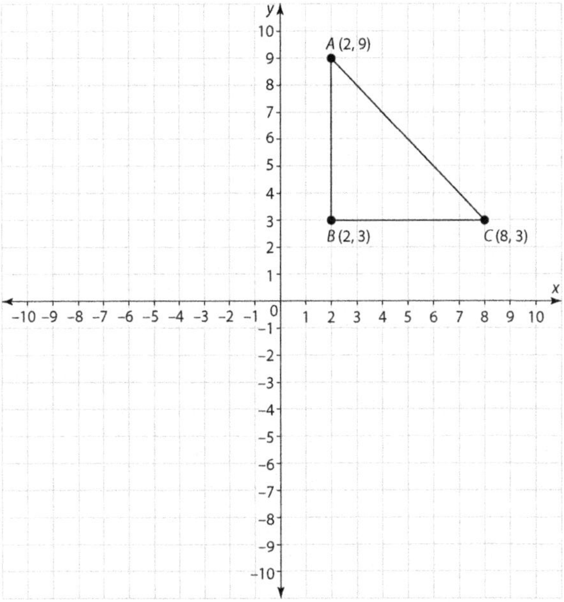

Which set of coordinates are the images, A', B', and C', of the vertices, A, B, and C, reflected over the y-axis?

A. $A' = (-2, -9)$, $B' = (-2, -3)$, and $C' = (-8, -3)$

B. $A' = (-2, 9)$, $B' = (-2, 3)$, and $C' = (-8, 3)$

C. $A' = (2, -9)$, $B' = (-2, -3)$, and $C' = (8, -3)$

D. $A' = (2, -9)$, $B' = (2, -3)$, and $C' = (8, -3)$

13. Rectangle *ABCD* is shown below.

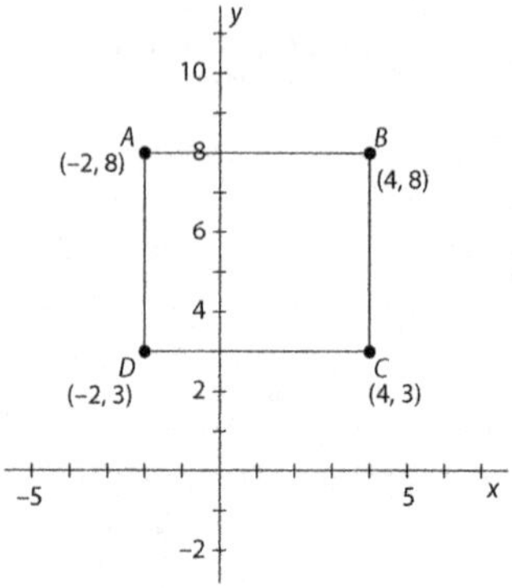

What is the perimeter of rectangle *ABCD*?

A. 11 units
B. 22 units
C. 30 units
D. 60 units

14. Evaluate: $7 + 3(4^2) - 8$

A. 23
B. 32
C. 47
D. 152

15. Evaluate the expression $10a - 2.2b$ when $a = 7.5$ and $b = 8$.

A. 57.4
B. 63.5
C. 92.6
D. 96.5

16. Which pair of expressions is equivalent?

A. $8x + 16y$ and $8(x + 2y)$
B. $2x + 6y$ and $2(x + 6y)$
C. $8(4xy)$ and $12xy$
D. $5x + 5y$ and $10xy$

17. The lake level rose 1.69 feet from March to June. The lake level in June was 628 feet above sea level. What was the lake level in March?

A. 625.31 feet
B. 626.31 feet
C. 628.69 feet
D. 629.69 feet

18. Three people split the cost of a taxi from the airport to a hotel. If each person paid $10.74, what was the cost for the taxi?

A. $30.12
B. $31.12
C. $32.22
D. $35.80

19. The table shows c, the cost in dollars, to rent n streaming movies. Write an equation that models the relationship between c and n.

Number of Movies *(n)*	**Cost in Dollars *(c)***
4	15.96
7	27.93
9	35.91

A. $c = n + 3.99$
B. $n = 3.99c$
C. $c = \frac{1}{3.99}n$
D. $c = 3.99n$

20. Determine the area of the given figure.

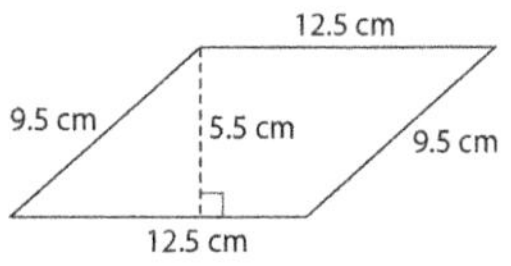

A. 34.375 cm^2
B. 44.0 cm^2
C. 68.75 cm^2
D. 118.75 cm^2

21. The figure shown has vertices *A* (5.5, 2.5), *B* (9.5, 2.5), *C* (9.5, –6.5), *D* (7.5, –6.5), *E* (7.5, –2.5), and *F* (5.5, –2.5). Determine its perimeter.

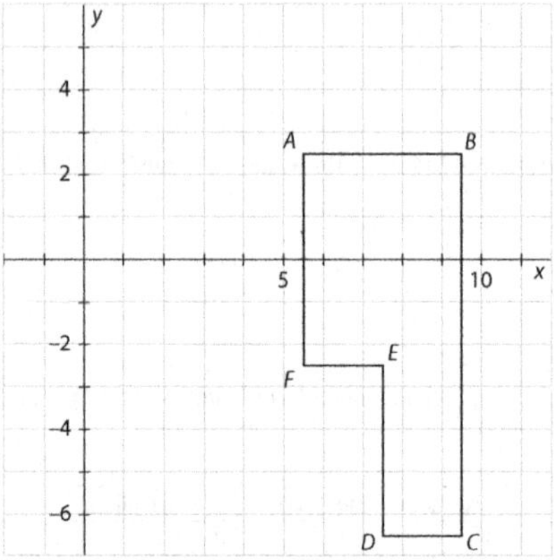

A. 24.5 units
B. 25.0 units
C. 25.5 units
D. 26.0 units

22. The triangle shown has vertices *A* (2, 0), *B* (4, 5), and *C* (4, –2). Determine its area.

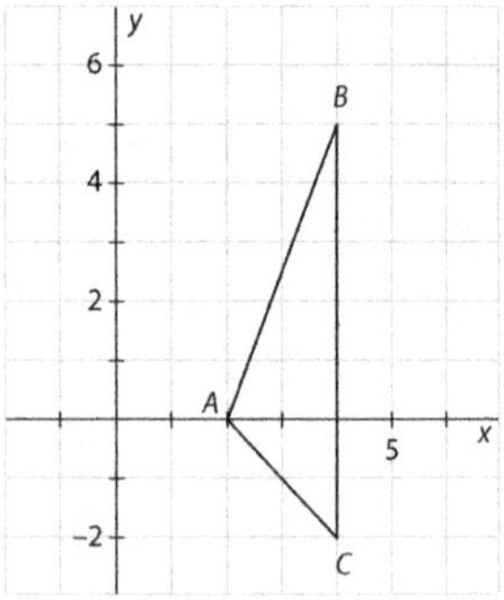

A. 7 units^2
B. 8 units^2
C. 14 units^2
D. 21 units^2

23. The weights in pounds of five sixth-grade boys are 80, 95, 84, 100, and 96. What is the mean weight of the five boys?

A. 84 pounds
B. 91 pounds
C. 95 pounds
D. 96 pounds

24. The dot plot shows the number of minutes 20 customers waited in line one morning at a restaurant.

Determine the range for the number of minutes waited in line by the 20 customers.

A. 15
B. 20
C. 25
D. 30

25. A box plot for the mathematics test scores of 10 sixth graders is shown below.

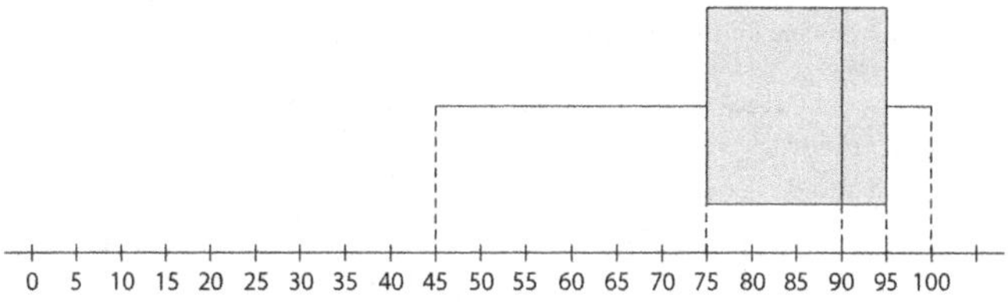

Determine the interquartile range for the students' test scores.

A. 20
B. 25
C. 30
D. 55

Directions: For questions 26–35, enter your answer in the answer box below the question. Enter the exact answer unless you are told to round your answer.

26. Nevaeh can type 240 words in 4 minutes. At this rate, how many words can Nevaeh type in 10 minutes?

[ ] words

27. Manolo is deciding whether to purchase a touch screen laptop computer for \$450.00 or to purchase a tablet priced at 0.71 as much as the laptop. What is the price of the tablet?

\$[ ]

28. When a number is multiplied by 12.5, the product is 230. What is the number?

[ ]

29. Determine the volume of the right rectangular prism shown.

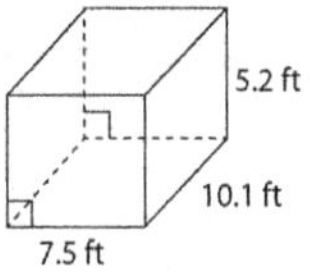

[ ] ft^3

30. What is the median score for the data shown in the dot plot below?

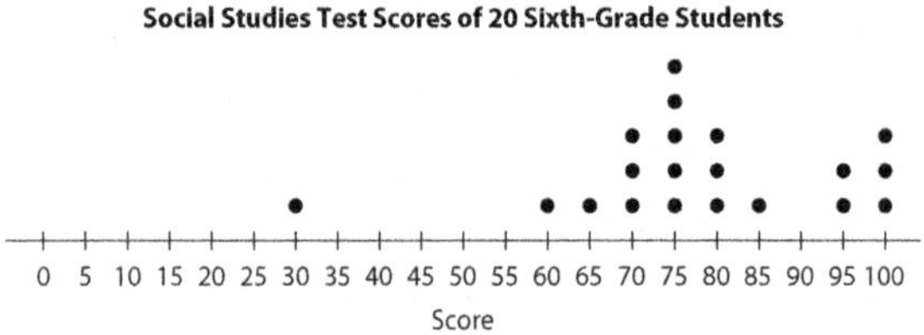

[ ]

31. Evaluate: $5^3 + 7 \times 2$

32. Compute: $3.42 + 2.875 + 5 + 1.6$

33. The coordinates of point A are $(-5, 10)$. The coordinates of point B are $(-5, -6)$. What is the distance between points A and B?

units

34. Solve: $4x = \frac{2}{5}$

35. In the diagram below, the sum of the measures of angles x and y is 90°. If the measure of $\angle y$ is 51°, what is the measure of $\angle x$ in degrees?

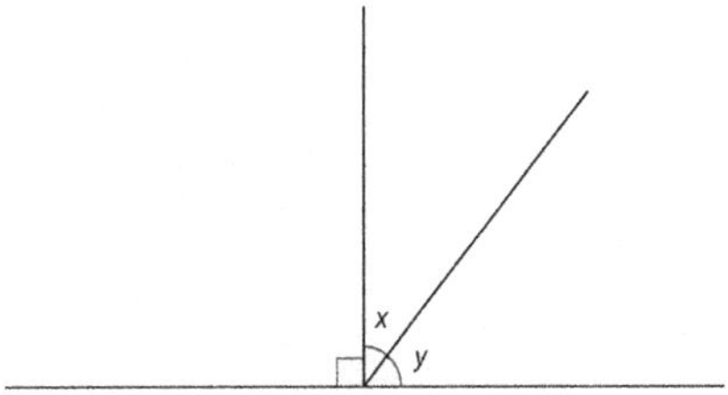

(figure not drawn to scale)

°

Answer Key

1. C
2. C
3. B
4. A
5. A
6. B
7. D
8. C
9. A
10. B
11. D
12. B
13. B
14. C
15. A
16. A
17. B
18. C
19. D
20. C
21. D
22. A
23. B
24. B
25. A
26. 600 words
27. $319.50
28. 18.4
29. 393.9 ft^3
30. 75
31. 139
32. 12.895
33. 16 units
34. $\frac{1}{10}$
35. 39°

Answer Explanations

1. **C.** Choice C is the correct response. Gallons and quarts are not the same units. So first, change gallons to quarts. One gallon equals 4 quarts. So, 2 gallons equals 8 quarts. Thus, the ratio of 2 gallons to 10 quarts equals the ratio of 8 quarts to 10 quarts. This ratio is $\frac{8 \text{ qt}}{10 \text{ qt}} = \frac{8 \cancel{\text{qt}}}{10 \cancel{\text{qt}}} = \frac{8 \div 2}{10 \div 2} = \frac{4}{5}$ or 4:5, Choice C.

Tip: Always check units to make sure they will cancel when you simplify a ratio.

2. **C.** Choice C is the correct response. The value of the ratio 12 to 18 is $\frac{12 \div 6}{18 \div 6} = \frac{2}{3}$. Check the answer choices to find which ratio is equivalent to $\frac{2}{3}$.

   Checking A: The value of the ratio 18 to 24 is $\frac{18 \div 6}{24 \div 6} = \frac{3}{4}$ (not equivalent).

   Checking B: The value of the ratio 12 to 30 is $\frac{12 \div 6}{30 \div 6} = \frac{2}{5}$ (not equivalent).

   Checking C: The value of the ratio 8 to 12 is $\frac{8 \div 4}{12 \div 4} = \frac{2}{3}$ (equivalent).

   Because you know C is the correct response, you can go on to the next question. Here is the check for D for your information.

   Checking D: The value of the ratio 4 to 9 is $\frac{4}{9}$ (not equivalent).

3. **B.** Choice B is the correct response. The number of cups of flour needed is $\frac{3}{20}$ times the number of pancakes produced. Therefore, making 50 pancakes would require $\frac{3}{20} \times 50 = \frac{3}{{}_{2}\cancel{20}} \times \frac{\cancel{50}^{5}}{1} = \frac{15}{2} = 7\frac{1}{2}$ cups of flour, Choice B.

4. **A.** Choice A is the correct response. Let G = the number of girls in the club, and B = the number of boys in the club. The number of girls is always $\frac{8}{5}$ times the number of boys. Using variable names, express this relationship as G is $\frac{8}{5}$ times B. Use the equation $G = \frac{8}{5}B$ to answer the question.

$$G = \frac{8}{5}B = \frac{8}{5} \times 15 = \frac{8}{{}_{1}\cancel{5}} \times \frac{\cancel{15}^{3}}{1} = \frac{24}{1} = 24$$

   There are 24 girls in the club, Choice A.

5. **A.** Choice A is the correct response. The percent is $5\% = \frac{5}{100}$. The part is 15 grams. The whole is the number that makes the two ratios, $\frac{5}{100}$ and $\frac{15}{\text{whole}}$, equivalent. Think: *"I want to get from 5 to 15. Because 15 is greater than 5, I will need to find a number to multiply 5 by to get 15. The number I need is 3."*

   Now multiply both the numerator and denominator of $\frac{5}{100}$ by 3.

   $$\frac{5 \times 3}{100 \times 3} = \frac{15}{300}$$

   Therefore, $\frac{15}{300} = \frac{15}{\text{whole}}$, which means the whole is 300.

   The mixture weighs 300 grams, Choice A.

6. **B.** Choice B is the correct response. Determine how many $\frac{2}{3}$ cups are in $2\frac{2}{3}$ cups. To find the answer, divide $2\frac{2}{3}$ by $\frac{2}{3}$.

   $$2\frac{2}{3} \div \frac{2}{3} = \frac{8}{3} \div \frac{2}{3} = \frac{8}{2} = 4$$

   There are 4 servings, Choice B.

7. **D.** Choice D is the correct response.

   $$0.005\overline{)2.5} = 0.005\overline{)2.500} = 5\overline{)\ 2500.}$$

   $$\begin{array}{r} 500. \\ 5\overline{)\ 2500.} \\ \underline{-25} \\ 0 \\ \underline{-0} \\ 0 \\ \underline{0} \\ 0 \end{array}$$

8. **C.** Choice C is the correct response. Solve this problem by finding the LCM of 4 and 6. Start by listing multiples of 6→ 6, 12. Stop because 12 = 3 × 4, a multiple of 4. So, the LCM of 4 and 6 is 12.

   Brenton and Jaomi will meet at the starting point again in 12 minutes, Choice C.

9. **A.** Choice A is the correct response. The number –5.25 is halfway between –5.30 and –5.20. So, it lies to the left of –5.2 (Choice A) and to the right of –5.3, –5.4, and –5.5.

10. **B.** Choice B is the correct response. Check each answer choice to determine which statement is true.

Checking A: $|-8.41| = 8.41$ and $|-14.15| = 14.15$. So, A is false because $8.41 < 14.15$. Think: "*–14.15 is farther from zero than –8.41 is. So, the absolute value of –14.15 is greater than the absolute value of –8.41.*" Choice A is false.

Checking B: $|-10.62| = 10.62$ and $|9.5| = 9.5$. So, B is true, because $10.62 > 9.5$. Think: "*–10.62 is farther from zero than 9.5 is. So, the absolute value of –10.62 is greater than the absolute value of 9.5.*" Choice B is true.

Because you know B is the correct response, you can go on to the next question. Here are the checks for C and D for your information.

Checking C: Think: "*–200 is farther from zero than 100 is. So, the absolute value of –200 is greater than the absolute value of 100.*" Choice C is false.

Checking D: Think: "*$5\frac{1}{2}$ is farther from zero than $-5\frac{1}{4}$ is. So, the absolute value of $5\frac{1}{2}$ is greater than the absolute value of $-5\frac{1}{4}$.*" Choice D is false.

11. **D.** Choice D is the correct response. Change the numbers to decimals.

4.100, 4.111 (rounded), 4.091 (rounded), 4.000, 4.110

Put the decimals in order.

4.000, 4.091, 4.100, 4.110, 4.111

Replace with the original numbers.

$4,\ 4\frac{1}{11},\ 4.1,\ 4.11,\ 4.\overline{1}$, Choice D.

12. **B.** Choice B is the correct response. The point $(-a, b)$ is the image of (a, b) when (a, b) is reflected across the y-axis. The image of A (2, 9) is A' (–2, 9), of B (2, 3) is B' (–2, 3), and of C (8, 3) is C' (–8, 3).

13. **B.** Choice B is the correct response. The perimeter of rectangle $ABCD = AB + BC + CD + AD$.

$AB = CD = |4| + |-2| = 4 + 2 = 6$; $BC = AD = |8| - |3| = 8 - 3 = 5$;
perimeter = 6 + 6 + 5 + 5 = 22 units

14. **C.** Choice C is the correct response. Follow the order of operations.

$$\begin{aligned} 7 + 3(4^2) - 8 &= 7 + 3(16) - 8 \\ &= 7 + 48 - 8 \\ &= 47, \text{ Choice C} \end{aligned}$$

15. **A.** Choice A is the correct response.

$$\begin{aligned} 10a - 2.2b &= 10(a) - 2.2(b) \\ &= 10(7.5) - 2.2(8) \\ &= 75 - 17.6 \\ &= 57.4 \end{aligned}$$

16. **A.** Choice A is the correct response. Check the answer choices.

    Checking A: $8x + 16y = 8(x) + 8(2y) = 8(x + 2y)$ (by the distributive property)
    Because you know A is the correct response, you can go on to the next question. Here are the checks for B, C, and D for your information.
    Checking B: $2x + 6y$ and $2(x + 6y)$ are not equivalent. If $x = 2$ and $y = 3$, $2x + 6y = 2 \cdot 2 + 6 \cdot 3 = 4 + 18 = 22$, but $2(x + 6y) = 2(2 + 6 \cdot 3) = 2(2 + 18) = 2(20) = 40$.
    Checking C: $8(4xy)$ and $12xy$ are not equivalent. If $x = 2$ and $y = 3$, $8(4xy) = 8(4 \cdot 2 \cdot 3) = 8(24) = 192$, but $12xy = 12 \cdot 2 \cdot 3 = 72$.
    Checking D: $5x + 5y$ and $10xy$ are not equivalent. If $x = 1$ and $y = 2$, $5x + 5y = 5 \cdot 1 + 5 \cdot 2 = 5 + 10 = 15$, but $10xy = 10 \cdot 1 \cdot 2 = 20$.

17. **B.** Choice B is the correct response. Let m = the March lake level in feet above sea level. Write an equation that represents the facts given in the problem.

    $$m + 1.69 \text{ feet} = 628 \text{ feet}$$

    Solve the equation. For convenience, omit the units while solving.

    $$\begin{aligned} m + 1.69 &= 628 \\ m + 1.69 - 1.69 &= 628 - 1.69 \\ m &= 626.31 \end{aligned}$$

    The lake level in March was 626.31 feet above sea level, Choice B.

18. **C.** Choice C is the correct response. Let c = cost of the taxi. Write an equation that represents the facts given in the problem.

    $$\frac{c}{3} = \$10.74$$

    Solve the equation.

    $$\begin{aligned} \frac{c}{3} &= \$10.74 \\ 3 \cdot \frac{c}{3} &= 3 \cdot \$10.74 \\ \frac{\cancel{3}^1}{1} \cdot \frac{c}{{}_1\cancel{3}} &= \$32.22 \\ c &= \$32.22 \end{aligned}$$

    The cost of the taxi was \$32.22, Choice C.

19. **D.** Choice D is the correct response. Determine the ratio of c to n.

    $$\frac{15.96}{4} = \frac{3.99}{1}, \ \frac{27.93}{7} = \frac{3.99}{1}, \text{ and } \frac{35.91}{9} = \frac{3.99}{1}$$

    So, c, the cost in dollars, is 3.99 times n, the number of movies. The equation is $c = 3.99n$, Choice D.

20. **C.** Choice C is the correct response. The figure has opposite sides congruent, so it is a parallelogram. The area, A, of a parallelogram is $A = bh$, where b is a base of the parallelogram and h is the height drawn to that base. Choose the side of length 12.5 centimeters as the base because you are given 5.5 centimeters as the height to that base.

$$A = bh = (12.5 \text{ cm})(5.5 \text{ cm}) = 68.75 \text{ cm}^2$$

21. **D.** Choice D is the correct response. The perimeter, P, of the figure is the sum of the lengths of its six sides. That is, $P = AB + BC + DC + ED + FE + AF$. The line segments composing the figure are either horizontal or vertical lines.

$\overline{AB}$ is a horizontal line. Both of its endpoints, A and B, are in the same quadrant. So, $AB =$ $|9.5| - |5.5| = 9.5 - 5.5 = 4.0$ units.

$\overline{BC}$ is a vertical line. Its endpoints, B and C, are in different quadrants. So, $BC = |2.5| + |-6.5| =$ $2.5 + 6.5 = 9.0$ units.

$\overline{DC}$ is a horizontal line. Both of its endpoints, C and D, are in the same quadrant. So, $DC =$ $|9.5| - |7.5| = 9.5 - 7.5 = 2.0$ units.

$\overline{ED}$ is a vertical line. Both of its endpoints, D and E, are in the same quadrant. So, $ED =$ $|-6.5| - |-2.5| = 6.5 - 2.5 = 4.0$ units.

$\overline{FE}$ is a horizontal line. Both of its endpoints, E and F, are in the same quadrant. So, $FE =$ $|7.5| - |5.5| = 7.5 - 5.5 = 2.0$ units.

$\overline{AF}$ is a vertical line. Its endpoints, A and F, are in different quadrants. So, $AF = |2.5| + |-2.5| =$ $2.5 + 2.5 = 5.0$ units.

So, $P = AB + BC + DC + ED + FE + AF =$ 4.0 units + 9.0 units + 2.0 units + 4.0 units + 2.0 units + 5.0 units = 26.0 units.

The figure's perimeter is 26.0 units, Choice D.

22. **A.** Choice A is the correct response. Enclose the triangle in a rectangle.

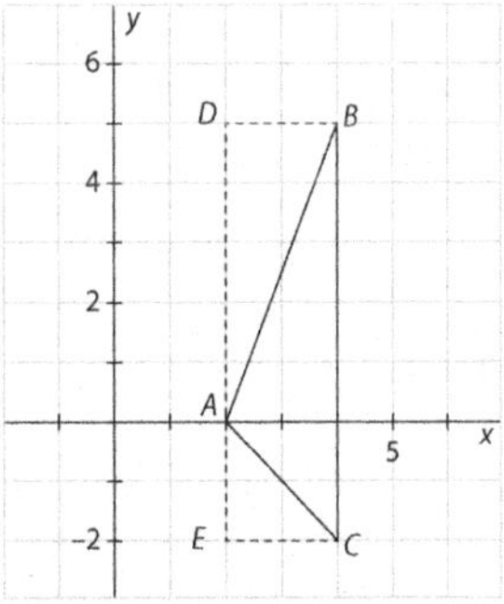

The area, A, of triangle ABC is the area of rectangle $DBCE$ minus the areas of right triangles ADB and AEC. In this figure, you can find the lengths of the sides of the figures by counting.

$$A = (7 \text{ units})(2 \text{ units}) - \frac{1}{2}(2 \text{ units})(5 \text{ units}) - \frac{1}{2}(2 \text{ units})(2 \text{ units})$$
$$= 14 \text{ units}^2 - 5 \text{ units}^2 - 2 \text{ units}^2$$
$$= 7 \text{ units}^2$$

Triangle ABC has area 7 units^2, Choice A.

23. **B.** Choice B is the correct response.

    Step 1. Sum the weights (omitting units for convenience).

    $$80 + 95 + 84 + 100 + 96 = 455$$

    Step 2. Divide by 5.

    $$455 \div 5 = 91$$

    The mean weight of the boys is 91 pounds, Choice B.

24. **B.** Choice B is the correct response.

    $$\text{range} = \text{maximum value} - \text{minimum value} = 25 - 5 = 20$$

25. **A.** Choice A is the correct response. The interquartile range (IQR) is the difference between the upper and lower quartiles of the students' test scores. The box plot shows the upper quartile is 95 and the lower quartile is 75.

    $$\text{IQR} = 95 - 75 = 20$$

    The interquartile range for the students' test scores is 20, Choice A.

26. **600 words**

    Step 1. Compute the rate for 1 minute.

    $$\frac{240 \text{ words}}{4 \text{ minutes}} = \frac{(240 \div 4) \text{ words}}{(4 \div 4) \text{ minutes}} = \frac{60 \text{ words}}{1 \text{ minute}}$$

    Step 2. Multiply the rate per minute by 10 minutes.

    $$\frac{60 \text{ words}}{\text{minute}} \times 10 \text{ minutes} = \frac{60 \text{ words}}{1 \cancel{\text{minute}}} \times \frac{10 \cancel{\text{minutes}}}{1} = \frac{600 \text{ words}}{1} = 600 \text{ words}$$

    Nevaeh can type 600 words in 10 minutes.

27. **$319.50**

$$\begin{array}{r} 450.00 \\ \times\ 0.71 \\ \hline 4\,5000 \\ 315\,000 \\ \hline 319.5000 \end{array}$$

The price of the tablet is $319.50.

28. **18.4**

Let n = the number. Write an equation that represents the facts given in the problem.

$$12.5n = 230$$

Solve the equation.

$$\begin{aligned} 12.5n &= 230 \\ \frac{12.5n}{12.5} &= \frac{230}{12.5} \\ \frac{{}^{1}\cancel{12.5}n}{{}_{1}\cancel{12.5}} &= \frac{\cancel{230}^{18.4}}{\cancel{12.5}_{1}} \\ 1n &= 18.4 \\ n &= 18.4 \end{aligned}$$

The number is 18.4.

29. **393.9 ft^3**

$$\text{Volume} = lwh = (7.5\text{ ft})(10.1\text{ ft})(5.2\text{ ft}) = 393.9\text{ ft}^3$$

30. **75**

The dot plot shows 20 scores. The number of data values is even, so the median is the average of the two middle values. Thus, the median is the average of the 10th and 11th data values, both of which are 75. The median score is 75.

31. **139**

$$\begin{aligned} 5^3 + 7 \times 2 &= 125 + 7 \times 2 \\ &= 125 + 14 \\ &= 139 \end{aligned}$$

32. **12.895**

$$3.42 + 2.875 + 5 + 1.6.$$

$$\begin{array}{r} 3.420 \\ 2.875 \\ 5.000 \\ +1.600 \\ \hline 12.895 \end{array}$$

33. **16 units**

The points *A* and *B* have the same *x*-coordinate, so they lie on the same vertical line. The *y*-coordinates have different signs. To find the distance between the two points, add the absolute values of the *y*-coordinates.

$$|10| + |-6| = 10 + 6 = 16$$

The distance is 16 units.

34. $\mathbf{\frac{1}{10}}$

$$4x = \frac{2}{5}$$

$$\frac{\cancel{4}x}{\cancel{4}} = \frac{2}{5} \div 4$$

$$\frac{{}^{1}\cancel{4}x}{\cancel{4}} = \frac{{}^{1}\cancel{2}}{5} \times \frac{1}{\cancel{4}_{2}}$$

$$1x = \frac{1}{10}$$

$$x = \frac{1}{10}$$

35. **39°**

Write an equation that represents the facts given in the problem.

$$x + y = 90°$$

Substitute 51° for y.

$$x + 51° = 90°$$

Solve the equation (omit the units for convenience).

$$x + 51 = 90$$

$$x + 51 - 51 = 90 - 51$$

$$x = 39$$

The measure of $\angle x$ is 39°.

Appendix: Measurement Conversions

U.S. Customary Units	Conversion
Length	
Inch (in)	$1\text{ in} = \frac{1}{12}\text{ ft}$
Foot (ft)	1 ft = 12 in $1\text{ ft} = \frac{1}{3}\text{ yd}$
Yard (yd)	1 yd = 36 in 1 yd = 3 ft
Mile (mi)	1 mi = 5,280 ft 1 mi = 1,760 yd
Weight	
Pound (lb)	1 lb = 16 oz
Ton (T)	1 T = 2,000 lb
Capacity	
Fluid ounce (fl oz)	$1\text{ fl oz} = \frac{1}{8}\text{ c}$
Cup (c)	1 c = 8 fl oz
Pint (pt)	1 pt = 2 c
Quart (qt)	1 qt = 32 fl oz 1 qt = 4 c 1 qt = 2 pt $1\text{ qt} = \frac{1}{4}\text{ gal}$
Gallon (gal)	1 gal = 128 fl oz 1 gal = 16 c 1 gal = 8 pt 1 gal = 4 qt

Metric Units	Conversion
Length	
Millimeter (mm)	$1\text{ mm} = \frac{1}{10}\text{ cm}$ $1\text{ mm} = \frac{1}{1000}\text{ m}$
Centimeter (cm)	1 cm = 10 mm $1\text{ cm} = \frac{1}{100}\text{ m}$
Meter (m)	1 m = 1000 mm 1 m = 100 cm $1\text{ m} = \frac{1}{1000}\text{ km}$
Kilometer (km)	1 km = 1000 m
Mass	
Milligram (mg)	$1\text{ mg} = \frac{1}{1000}\text{ g}$
Gram (g)	1 g = 1000 mg $1\text{ g} = \frac{1}{1000}\text{ kg}$
Kilogram (kg)	1 kg = 1000 g
Capacity	
Milliliter (mL)	$1\text{ mL} = \frac{1}{1000}\text{ L}$
Liter (L)	1 L = 1000 mL

Time	Conversion
Second (s)	$1\text{ s} = \frac{1}{60}\text{ min}$ $1\text{ s} = \frac{1}{3{,}600}\text{ hr}$
Minute (min)	1 min = 60 s $1\text{ min} = \frac{1}{60}\text{ hr}$
Hour (hr)	1 hr = 3,600 s 1 hr = 60 min $1\text{ hr} = \frac{1}{24}\text{ d}$
Day (d)	1 d = 24 hr
Week (wk)	1 wk = 7 d
Year (yr)	1 yr = 365 d 1 yr = 52 wk

Made in the USA
Monee, IL
04 November 2026

67379459R00125